IPP Studies in the Frontiers of China's Public Policy

Series Editor
Feng Zhang, South China University of Technology, Guangzhou, China

IPP Studies in the Frontiers of China's Public Policy combines original research and theoretical innovation to provide fresh insights into the fast-changing landscape of China's public policy. Books in the series, written by scholars based inside China or commissioned by the IPP, draw on local Chinese experiences to generate empirical findings and theoretical insights. The field of China's public policy is broadly defined to include all aspects of the country's social, economic, technological and foreign policies. The series encourage interdisciplinary approaches to public policy, with a special but not exclusive focus on the study of south China, especially the Guangdong, Hong Kong, Macau Greater Bay Area, one of the most vibrant regions of growth and innovation in China.

Yuhao Jiang · Kai Jia

Transformation of Platform Governance in China

The Politics of Technology Routes

palgrave
macmillan

Yuhao Jiang
Institute of Public Policy
South China University
of Technology
Guangzhou, China

Kai Jia
School of International and Public
Affairs
Shanghai Jiao Tong University
Shanghai, China

IPP Studies in the Frontiers of China's Public Policy
ISBN 978-981-99-6455-0   ISBN 978-981-99-6456-7 (eBook)
https://doi.org/10.1007/978-981-99-6456-7

Cover illustration: Nikada

This Palgrave Macmillan imprint is published by the registered company Springer Nature Singapore Pte Ltd.
The registered company address is: 152 Beach Road, #21-01/04 Gateway East, Singapore 189721, Singapore

Paper in this product is recyclable.

Acknowledgments

This book represents our preliminary thoughts on the rapidly developing new information and communication technologies (new-ICT), as well as the digital platforms built upon them as new modes of production and business organization. Among the two authors, Jiang Yuhao's research interests focus on political theory and transformation of China's political system. In this book, his contributions revolve around an impact of China's political system and public policies on the development of digital platforms. Due to his professional background in electronic engineering, Jia Kai's research interests aim to find the intersection between digital technology and political theory. Although the research interests of the two authors differ, their shared passion lies in exploring the multiple possibilities of digital transformation in the future. This shared passion owes much to the theoretical writings of Professor Roberto Mangabeira Unger from Harvard Law School and Professor Charles F. Sabel from Columbia Law School. Professor Unger's book *The Knowledge Economy* (Verso, 2019) and Professor Sabel's recent research papers on the concept of 'Experimentalism Governance' serve as the starting point for this book. In particular, Professor Unger's visit to Tsinghua University from 25 to 29 October 2017, provided us with invaluable face-to-face exchanges. From the writings of Unger and Sabel, we not only gained knowledge but also acquired wisdom, perspective and courage to critically interpret the status quo.

The reason why we could build up the relatively rich knowledge is entirely due to the guidance and mentorship of our common teacher, Professor Cui Zhiyuan of Tsinghua University School of Public Policy and Management. It is worth noted that the core concept of this book, 'The Politics of Technology Routes', was formed under the long-term guidance of Professor Cui. We are grateful to have experienced the joy of scholarly reading and thinking in this dynamic academic community led by Professor Cui. In addition, Jiang Yuhao wishes to thank Professor Zheng Yongnian, from the Institute of International Affairs at the Chinese University of Hong Kong (Shenzhen), for his international strategic insights and thought-provoking analysis of Chinese development issues, which have frequently provided surprising inspiration.

The writing of this book has taken several years. During this time, we have received help from many friends, including editors of relevant Chinese and English journals who have published related articles, entrepreneurs, and government officials who have assisted us in our research, and especially our friends who have mutually inspired us in our academic thinking. We cannot list each of their names individually, but we express our gratitude to all of them. The two authors would like to thank their respective institutions, the Institute of Public Policy at South China University of Technology (*IPP*) and the School of Public Administration at the University of Electronic Science and Technology of China.

Jiang Yuhao would like to extend a special thanks to his wife, Mrs. Liu Zhixiu. She has put in a tremendous amount of effort to help bring this book to its final completion and has provided invaluable suggestions based on her professional knowledge and practical work experience. Jiang Yuhao wishes to dedicate this work to Mrs. Liu Zhixiu and their two lovely and intelligent daughters, Zhuolin and Zhuoxi.

Jia Kai would like to express his heartfelt gratitude to his wife, Mrs. Hu Wenting. Their mutual support and encouragement have been crucial in their shared academic explorations. It is not about having the same knowledge foundation, but rather the curiosity they both hold toward the world and history. Jia Kai wishes to dedicate this book to Mrs. Hu Wenting and their daughter, Jia Xun, as a token of appreciation for their companionship and enduring support on this shared journey.

Contents

List of Figures

List of Tables

CHAPTER 1

Introduction: Technological Determinism, Insular Vanguardism and Possibility of Platform Transformation

Techno-liberalism and Techno-nationalism: Opposing Categories?

The Internet is the core invention and application of the information technology revolution that began in the 1970s. Today, the new generation of information technology, such as mobile Internet, cloud computing, Internet of Things, big data and artificial intelligence, is accelerating, and the information technology industry is entering a new round of rapid growth. General Electric (GE) predicts the Industrial Internet will contribute US$15 trillion to global GDP by 2030. This trend has led to widespread recognition among political elites worldwide that promoting the integration of new-generation information technology with technological, economic and social development can be a significant strategy for future competition. However, it is unclear whether the rapid iteration of new information technologies will add to the well-being of various countries' inhabitants or merely increase each side's leverage in geopolitical competition and conflict.

On 14 September 1987, the Chinese Academic Network (CANET), an international networking project implemented by the Peking Institute of Computer Application Technology, sent the first Chinese email to the Karlsruhe Institute of Technology in Germany: 'Across the Great Wall we can reach every corner in the world. Across the Great Wall we

Y. Jiang and K. Jia, *Transformation of Platform Governance in China*, IPP Studies in the Frontiers of China's Public Policy,
https://doi.org/10.1007/978-981-99-6456-7_1

can reach every corner in the world,' unveiling the first Chinese access to the Internet.[1] The Chinese scientists and engineers who were encouraged by the success of their access to the global network did not realize that being aligned with the frontiers of new information technology would soon bring China face to face with a more complex and diverse future. While the new information technology has given China more opportunities for scientific and technological development and economic growth, it has also created many new and more complex challenges for the country's governance.

Today, China's Internet infrastructure construction has ranked among the world's foremost, with various indicators such as network coverage, number of Internet users and frequency of Internet access exceeding the world average: as of December 2018, the size of China's Internet users was 829 million, with an Internet penetration rate of 59.6%; the size of mobile phone Internet users reached 817 million; the proportion of Internet users accessing to the Internet by using mobile phones reached 98.6%; and mobile phone Internet access has become. The number of rural Internet users was 222 million, accounting for 26.7% of the total number of Internet users; the average weekly Internet access time of Internet users was 27.6 hours. Internet users are mainly young and middle-aged, and there is a trend of continuous penetration to the middle and senior age groups: at the end of 2018, the group aged 10–39 accounted for 67.8% of the overall Internet users, of which the proportion of Internet users in the age group of 20–29 was the highest at 26.8%, the proportion of middle-aged Internet users aged 40–49 expanded from 13.2% at the end of 2017 to 15.6%, and the proportion of Internet users aged 50 and above increased from 10.5% at the end of 2017 to 12.5%.[2] With the help of the Internet, Internet users engage in various activities such as instant messaging, shopping, entertainment, financial management and education. The Internet economy accounted for more than 7% of the gross domestic product (GDP) in 2015, surpassing the

[1] There is a debate between 25 August 1986 and 14 September 1987 as to when the first email was sent in China. The more popular general statement is adopted here, according to the Internet Society of China, History of Internet Development in China (Chronology), https://www.isc.org.cn/ihf/info.php?cid=218. (Last accessed: 10 June 2022, same below).

[2] CNNIC, The 43rd Statistical Report on the Development Status of the Internet in China 2019, http://www.199it.com/archives/839540.html.

United States to become the world's number one, and is still increasing at an average annual rate of 30%.[3] The virtual world of the Internet has become an inseparable part of Chinese people's daily life.

The theoretical understanding of the opportunities and challenges posed by the rise of new information technologies in China under communist governance is a subject of endless debate in academic research. In fact, from the early days of the rise and popularity of the Internet, there has been much debate about whether new information technologies are empowering society (i.e., expanding the freedom of action and capabilities of social agents) or empowering the state (i.e., enhancing the state's ability to monitor and manage society). The controversy is not just about different perceptions of an 'objective' social phenomenon, but also about deeper concerns and political ideas about the future of human society. Nicholas Negroponte, a founder of the MIT Media Lab, published his classic book 'Digital Survival' in 1996, predicting the advent of a digital age of freedom, in which his vision of a free association of free people across borders and beyond the law, based on new information technologies, inspired countless talented young people around the world to join the Internet industry. However, in his preface to the 20th-anniversary edition of the book in Chinese translation in 2016, Negroponte, the 'prophet' of the cyber-information age, described the actual development in pessimistic terms: 'The real situation is: nationalism is on the rise; regulation is escalating; the gap between rich and poor is growing. The real situation is: nationalism is rampant; regulation is escalating; the divide between rich and poor is growing ... is simple: globalization has become localization, despite the rapid growth of the intellectual, economic and electronic backbone. Ubiquitous digitization has not brought about a cosmopolitan world.'[4] For former Techno-optimists/Techno-liberalists such as Negroponte, the subordination of the Internet to the regulation of the nation-state and its fragmentation by nation-state frontiers is a sign of the degeneration of the ideal of a free world, where new information technologies, while strengthening the regulatory capacity of the state, suppress the space for autonomous social action and the arena for

[3] Jianjun, Dai and Tian Jietang, *Research on the Regulation of Emerging Internet Industries*, Peking: China Development Press, 2017, p. 5.

[4] Negroponte, Nicholas, *A Special Preface to the 20th Anniversary Chinese Edition of Digital Survival*, translated by Fan, Haiyan, in Future World: http://www.51wxjz.com/100473.

the expression of liberal ideas.[5] Changing trends in geopolitics seem to prove that techno-nationalism does not work for the good of the world: the trade conflict between the US and China that has erupted since 2018, driven by radical techno-nationalists in the US, has led to the inability of a significant number of people in the Western world to imagine that the US and China can achieve, through cooperation in the field of production and normal and peaceful institutional competition, the integration of the industrial value chains of both countries and the improvement of their respective institutions.[6] At the same time, a strong critique of techno-liberalism remains strong: techno-liberalists embrace nothing more than a picture of elite-dominated globalization, and the ubiquitous worldwide worsening of the gap between rich and poor shows that the regulation of new information technologies by nation states should remain the order of the day.[7]

Here is a theoretical question worthy of analysis: Techno-liberalism (favoring technological autonomy and individual freedom) and techno-nationalism (favoring the enhancement of national regulatory capacity) are only a pair of opposing categories, and they cannot form a mutually reinforcing and interdependent relationship? When Xi Jinping attended the opening ceremony of the Second World Internet Conference, he made a speech in which he explicitly stressed the importance of 'promoting openness and cooperation' to promote changes in the global Internet governance system: 'If the world loves each other, it will be ruled, but if it crosses paths, it will be chaotic. To improve the global Internet governance system and maintain order in cyberspace, it is necessary to adhere to the concept of being in the same boat, mutual trust and mutual benefit, and abandon the old concept of zero-sum game and winner takes all. Countries should promote open cooperation on the Internet, enrich the connotation of openness, raise the level of openness, build more platforms for communication and cooperation, create more points of convergence of interests, growth points of cooperation and new

[5] Jack Goldsmith and Tim Wu, *Who Controls the Internet: Illusions of a Borderless World*, Oxford: Oxford University Press, 2006.

[6] Zheng Yongnian, China and the U.S.: A Clash of Techno-Nationalists, *The Washington Post*, May 15, 2018; Cui Zhiyuan, Greater Convergence More Likely than Decoupling, *China Daily*, October 9, 2019.

[7] Carl Benedikt Frey, *The Technology Trap: Capital, Labor, and Power in the Age of Automation*, Princeton & Oxford: Princeton University Press, 2019.

bright spots for win–win cooperation, and promote each other's complementary advantages and common development, so that more countries and people can ride the express train of the information age and share the fruits of Internet development.'[8] Whatever one's view of China, this statement by President Xi Jinping is commendable, and this thinking of advocating cooperative and shared governance of the Internet has been implemented to some extent in China's policy-making practices.[9] However, it is clear that this new pattern of open, shared, cooperative and win–win global network governance can only be promoted by using changes in the national governance system as the engine. We have argued in other works that the cyber regulatory system in the new era should be based on a profound understanding of the dialectical relationship between openness, sharing and win–win cooperation, so that techno-liberalism and techno-nationalism are organically integrated with the design of specific systems, and the capacity of the state to develop and apply new information technologies is elevated to 'the political capacity to promote shared freedom.'[10]

So, is it possible and how can China develop a 'political capacity for shared freedoms' in the practice of new IT governance? This book proposes two levels of reflection: first, at the institutional level, i.e., whether and how China's new information technology governance system can develop a regulatory approach conducive to shared freedoms? Second, at the technological level, can new information technologies and the platforms on which they are built be challenged and transformed in order to shape shared freedoms? In the existing theoretical literature, there is a

[8] Jinping Xi, Speech at the Opening Ceremony of the Second World Internet Conference (full text), in Xinhua: http://www.xinhuanet.com/politics/2015-12/16/c_1117481089.htm.

[9] In response to the rapid iteration of new information technologies, China's Ministry of Foreign Affairs publicly released its 'China's Position on Issues Relating to Global Digital Governance (Submission to the United Nations on the Development of a Global Digital Compact)' in April 2023, reflecting President Xi Jinping's thinking on cooperative governance of the Internet. In addition, the rapid rise of ChatGPT has sparked panic around the world, and countries have recently been formulating policies and ethical norms to regulate the research and development of AIGC. Compliance obligations of service providers.

[10] Jia Kai and Jiang Yuhao, *The Technology and Politics of Digital Governance*, Peking: Social Science Literature Press, forthcoming.

rich discussion of the first institutional dimension, but very little discussion of the second technical dimension. In this book, the neglect of the possibility of multiple Technology Routes in the theoretical literature suggests that a certain 'Technological Determinism' still dominates current research on the development of new information technologies and platform governance. Technological determinism leads to a failure to see the opportunities for convergence and innovation between techno-liberalism and techno-nationalism, and to abandon the exploration of new Technology Routes to change the status quo. The ultimate possibility of shared freedom.

The rapid development of digital and intelligent technologies, represented by the new generation of information and communication technologies, has driven dramatic economic and social changes while also plunging society into deep anxiety. In recent years, significant countries worldwide have launched various initiatives on national security, platform regulation, privacy protection, labor protection and so on, aiming to address the economic and socio-political risks inherent in the digital age. However, as Shoshana Zuboff argues, the most severe threat to the rapid rise of mega-platforms is the autonomy and self-governance of citizens.[11] The greatest crisis is that a significant number of people are at risk of being excluded from the rapid development of new technologies and are unable to participate in the process of technological progress. In their recent work, Roberto M. Unger and Amy Kapczynski have argued more profoundly that institutional innovation in the digital age should aim to stimulate the autonomous participation of ordinary citizens in technological progress and to change the institutional and technological conditions that allow the rapid expansion of mega-platforms.[12]

The new technological revolution was widely anticipated because it conceptually promised disruptive changes to established ways of production and life, advocating a break with routine and mechanistic repetition and introducing innovation into everyday habits and institutional arrangements. However, as will be discussed in more detail below, in practice, the invention and application of new technologies have primarily reinforced the increasingly rigid division of labor, siloed into areas where capital

[11] Shoshana Zuboff, *The Age of Surveillance Capitalism, The Fight for a Human Future at the New Frontier of Power*, New York: Public Affairs, 2019.

[12] Roberto M. Unger, *The Knowledge Economy*, London and New York: Verso, 2019; Amy Kapczynski, The Law of Informational Capitalism, 129 *Yale L. J.* 1460 (2020).

and technological capabilities were already concentrated, and consolidated the competitive position of giant firms with significant advantages in terms of economies of scale and centralized management, which calls for a profound rethink. Liberalists let this insular pioneering character of new technologies go unchecked; in their view, as the new information technology revolution develops further, more convenient and famous technological inventions will continue to emerge, and society as a whole will benefit from the continued progress of such technologies and the continued spread of new technological products; techno-nationalists exacerbate this insular pioneering character of new technologies, according to their techno-nationalists exacerbate this insular vanguardist character of new technologies, according to their belief that the promotion of the new information technology revolution and the embrace of new technological inventions are the ways to achieve a wealthy nation and a reliance on security and competitiveness in the face of geopolitical conflict. Both ideas are expressions of 'technological determinism': technology is a technology and human agency can only accelerate or slow down its development, but not change its direction. Challenging 'technological determinist' thinking is to break down the apparent opposition between the categories of techno-liberalism and techno-nationalism, and to open up the possibility of moving from an insular vanguardism to a more shared freedom of new technologies.

Insular Vanguardism

The current rapid development of new digital and intelligent technologies is, in R. M. Unger's terminology, a form of 'insular vanguardism': innovation activities and application practices are mainly confined to a few frontier regions and industries, and are not effectively and widely diffused. It needs to be more effectively and widely diffused and has profoundly transformed production methods and unleashed greater productivity.[13] Statistics from the United States show that the rapid development of new technologies has been accompanied by a significant decline in total factor productivity and labor productivity growth in recent years: from 2004 to 2014, the average annual growth rate of total factor productivity in the United States was 0.4%, the lowest in history since 1890; from 2005 to

[13] Roberto M. Unger, *The Knowledge Economy*, London and New York: Verso, 2019.

2016, the average annual growth rate of total labor productivity in the United States was 1.3%, less than 2% of the average annual growth rate in the previous decade (1995–2004). From 2005 to 2016, total labor productivity in the US grew at an average annual rate of 1.3%, less than half the 2.8% average annual growth rate of the previous decade (1995–2004). Data from the US Bureau of Labor Statistics further show that US manufacturing productivity grew at an average annual rate of 4.1% between 1989 and 2000, while between 2007 and 2014, manufacturing productivity, which absorbed the benefits of the IT revolution, fell to an average annual growth rate of 1.7%.[14]

In 1987, Robert M. Solow, winner of that year's Nobel Prize in Economics and founder of neoclassical growth theory, put forward the famous 'productivity paradox'—'computers are everywhere, except in productivity statistics'—questioning the minimal contribution of the IT industry to productivity.[15] Economist Robert Gordon's systematic research goes even further, pointing out that the scope and depth of impact of current digital, intelligent new technologies are minimal compared to the second Industrial Revolution, which transformed all significant aspects of production and lifestyles worldwide. So far, for example, smartphones have been used mainly for personal activities in the office; e-commerce accounts for too low a proportion of the economy as a whole; and the integration of digital technologies into the financial and banking sectors has not reduced companies' reliance on professionals such as financial accountants or offset the willingness of banks to expand their physical branches—in the words of Robert Gordon 'Each of the sectors discussed above (office, retail, finance, banking) underwent fundamental and revolutionary changes in the 1980s and 1990s. In the last 10 years, computer hardware, software and business methods have ossified into a set of slowly changing practices.'[16]

On the one hand, the technological innovations that most Western developed countries are focusing on have done little to hollow out their natural industries. Ordinary households at home who cannot afford to enter the high-tech and high-end service sectors have been unable to

[14] William B. Bonvillian and Peter L. Singer, *Advanced Manufacturing: The New American Innovation Policies*, Cambridge: The MIT Press, 2017.

[15] Robert Solow, We'd Better Watch Out, *New York Times Book Review*, July 12, 1987.

[16] Robert Gordon, *The Rise and Fall of American Growth*, translated by Zhang, Linshan et al., Peking: CITIC Publishing Group, 2018: 558.

share in the dividends of the new technological revolution, thus triggering a wave of anti-globalization and Neo-Luddism. On the other hand, as international economist Dani Rodirik has pointed out, in developing countries and regions involved in the formation of global value chains, as the effects of new technology-driven industrial development and exports have become increasingly evident, global value chains have neither expanded employment nor contributed to export-related technological and organizational benefits throughout the economy. Instead, the traditional comparative advantage of these countries and regions in terms of unskilled labor has been eroded by the labor-saving and resource-saving effects of new technologies.[17]

To a large extent, the prevailing institutional environment for scientific and technological research and development and technological innovation is centered on a rigid distinction between 'intellectual work' and 'manual work'; there are various laws and regulations governing the creation, innovation, diffusion, application and protection of new technologies The disconnect between technological innovation and the working practices and daily needs of the general population is exacerbated by the fact that there is a disproportionate incentive for technological research and development outside of productive labor practices. This can be recognized in the famous criticism of one of the pioneers of the new technological wave—Peter Thiel, founder of PayPal and the first external investor in Facebook—who said: 'We wanted a car that flies, and all we got was 140 characters.'

Marx severely criticized the rigid distinction mentioned in Volume I of Capital: 'The separation of the intellectual from the manual labor of the production process, the transformation of the intellect into the power of capital to dominate labor, is accomplished in the great industry based on machines. ...... Science, the immense forces of nature, and the mass labor of society are embodied in the machine system and constitute the power of the 'master'.'[18] In Marx's claim, the 'social use of

[17] Dani Rodirik, New Technologies, Global Value Chains, and the Developing Economies, NBER Working Paper Series, October 2018, http://www.nber.org/papers/w25164.

[18] Marx, Karl, Capital, vol. 1, in *Marx and Engels Collected Works*, vol. 5, translated by the Bureau for the Compilation of the Works of Marx, Engels, Lenin and Stalin of the Central Committee of the Communist Party of China, Beijing: People's Publishing House, 2009: 487.

the material means of production' should be achieved by appropriating the material means of production by the working class. However, this understanding, although highly theoretically insightful, can easily be interpreted in one-sided way as a revolutionary and subversive mechanism that can be pursued in one fell swoop, leaving everyday life experiences entirely behind. Irrespective of the ownership of the means of production, the key to ensuring that technological innovation, scientific research and labor practices are closely integrated is that the mass of workers and ordinary individuals can participate in technological progress on their own. In other words, considering the question of ownership alone will not help to eliminate the idea of a rigid system that separates 'intellectual work' from 'manual work.' This was neglected in the Soviet Union's technological development, which was mainly aimed at satisfying heavy industrial development plans and setting highly sophisticated research targets, resulting in a severe disconnect between scientific research, technological inventions, production processes and the needs of life, which prevented the outstanding achievements of scientific and technological progress from being reflected in the daily development of the economy and society. In the post-World War I era, the United States focused on increasing investment and support at the 'front end of innovation,' resulting in rapid progress in scientific and technological research and development and the emergence of a world innovation hub such as Silicon Valley. However, the result was that scientific and technological innovation became increasingly 'self-referential' under the combined effects of the financial system, venture capital mechanisms, intellectual property protection systems and other institutional conditions. It has become increasingly 'self-referential' to the development needs of traditional economic sectors such as manufacturing, fossil fuel energy, electricity grids, healthcare delivery systems, highway-intensive transport and input-intensive agriculture, with declining labor productivity, a widening gap between rich and poor, rising populist politics and, in short, an 'innovation-production' disconnect. The 'innovation-production' disconnect has led to severe economic and socio-political problems. In recent years, the federal government has been 'reinventing manufacturing,' and one of the most critical initiatives in its policy is how to 'get its research and development institutions to include advanced manufacturing in their research portfolios.'[19]

In any case, the rigid institutional concept that separates 'intellectual work' from 'manual work' is at the root of the 'insular Vanguardism' (Roberto M. Unger) character of new technologies. Under this philosophy, the actual needs of productive labor practices and the knowledge accumulated in the process denied the opportunity to be refined, fostered and developed, while the creation and application of new technologies that are difficult to integrate into more comprehensive production and living scenario risk further undermining labor creativity, which could lead to serious development imbalances and exacerbate social divisions.

Different Routes of Technological Development: Anti-technological Determinism

There have always been two different attitudes to technological development: optimism and pessimism, but these two opposing attitudes often fail to recognize a familiar blind spot. Simply put, whether optimistic or pessimistic, the perception of technology may be viewed statically from the outside of technology, as the result of applying of scientific principles, without considering the possibility of different Technology Routes in a given context, with different application prospects and socio-political consequences. David F. Noble, a historian of technology at MIT, is a profound critic of technological determinism: 'Technological determinism offers a simple explanation for things—especially troublesome things—and holds out the prospect of automatic and inevitable solutions. Ratifying the status quo as necessary at this stage of development, technological determinism absolves people of the responsibility to change it and weds them instead to the technological projections of those in command. Thus, if this ideology simplifies life, it also diminishes life, fostering compulsion and fatalism, on the one hand, and an extravagant, and unpredictable solution. Thus, if this ideology simplifies life, it also diminishes life, fostering compulsion and fatalism, on the one hand, and an extravagant, futuristic, faith in false promises, on the other.'[20] The

[19] William B. Bonvillian and Peter L. Singer, *Advanced Manufacturing: The New American Innovation Policies*, Cambridge: The MIT Press, 2017, p. 10.

[20] David F. Noble, *Forces of Production: A Social History of Industrial Automation*, New Brunswick and London: Transaction Publishers, 2011, Preface xiii.

existing Technology Route defines the relationship between technology, capital and labor, considering this hierarchy as 'the status quo,' which has created an insurmountable gap between 'intellectual labor' and 'manual labor.' Under such a situation, it is difficult for ordinary people, including most of the working population, to develop relatively autonomously. 'The choice is inescapable.'

Daron Acemoglu, a star economist at MIT, and his co-workers have recently launched a discussion on 'The Wrong Kind of AI,' warning about the mainstream AI Technology Route—a solution to AI that favors automation. The current mainstream AI technology change tends to move toward automation, allowing cheaper capital (machines) to replace a range of tasks previously performed, without significant productivity gains to increase the demand for labor. The consequences of this could be disastrous for income inequality and social cohesion. Acemoglu et al. demonstrate the reality of these dangers with their empirical study:

> We have previously studied the implications of industrial robots, one of the most important automation technologies. Industrial robots are not technologies aimed at increasing labor's productivity but are designed to automate tasks that production workers previously performed by on the factory áoor. The evidence is fairly clear that industries with more industrial robots are introduced experience declines in labor demand (especially for production workers) and increases in labor costs. The evidence is fairly clear that industries where more industrial robots are introduced experience declines in labor demand (especially for production workers) and sizable falls in their labor share. Furthermore, the declines in wages and employment fell much more heavily on workers from the lower half of the workforce. Furthermore, the declines in wages and employment fell much more heavily on workers from the lower half of the earnings distribution and those with less than a college degree, thus exacerbating inequality. All of this is industry-level data but also suggests productivity gains from robots.[21]

We can recognize the importance of opinions such as Asimoglu's when we look back at the social controversies arising from the development of technological innovation in the United States and its current reflections.

[21] Daron Acemoglu & Pascual Restrepo, The Wrong Kind of AI? Artificial Intelligence and The Future of Labor Demand, NBER Working Paper Series, http://www.nber.org/papers/w25682.

Out of the need to compete with the Soviet Union in science and technology after World War II, the US spawned an information technology revolution led by the Department of Defence agencies. For example, following the launch of the first human satellite by the Soviet Union in 1957, the US established the Defense Advanced Research Projects Agency to fund scientists at various universities and research institutions to freely pursue scientific research and translate the results, and also funded the Computer Science Division to provide early research support to start-ups, promote semiconductor research and human–computer interaction research, and lead to the early development of the Internet. Fairchild Semiconductor in Silicon Valley, which later played a decisive role in the history of semiconductor technology, grew to become a world leader in semiconductor research and development after its initial development was primarily due to military procurement. However, the focus of the information technology revolution was on producing a new generation of products rather than on improving the competitiveness of American manufacturing. As early as the late 1970s and early 1980s, 'the number of new jobs created by high technology will fall disappointingly short of those lost in manufacturing.'[22] The decline of manufacturing has become irreversible in the twenty-first century: in terms of employment, from 1965 to 2000, the number of manufacturing jobs in the US remained roughly 17 million, while from 2000 to 2010, the number of manufacturing jobs in the US remained roughly 1.5 million. Between 2000 and 2010, manufacturing employment plummeted by almost a third, with job losses of 70% and 50%, respectively, in low value-added sectors such as textiles and furniture, which are vulnerable to globalization; in terms of investment, overall business investment as a share of GDP has fallen by 3% since the 1980s as of 2012, and between 2000 and 2013, some 64,000 manufacturing plants closed, followed by only a modest recovery; US manufacturing output grew at an annual rate of just 0.5% from 2000 to 2007 and zero from 2007 to 2014, and during the financial crisis, manufacturing output fell sharply by 10.3% from 2007 to 2009, before entering the slowest US economic recovery in 60 years recovery period.[23] The

[22] David F. Noble, *Forces of Production: A Social History of Industrial Automation*, New Brunswick and London: Transaction Publishers, 2011, p. 348.

[23] William B. Bonvillian and Peter L. Singer, *Advanced Manufacturing: The New American Innovation Policies*, Cambridge: The MIT Press, 2017, pp. 52–56.

vigorous development of technological innovation divorced from manufacturing production practices has not helped much positively with the general weakening of the US industrial base. However, it has instead produced the kind of danger that urban development historian Lewis Mumford warned about back at the beginning of the second Industrial Revolution: 'For the drive toward automation, as Mumford has warned, steadily undermines the small-scale, decentralized, skill-based, versatile, and durable industrial For the drive toward automation, as Mumford has warned, steadily undermines the small-scale, decentralized, skill-based, versatile, and durable industrial infrastructure—one foundation, at least, of democracy and, as the surest carrier of the accumulated knowledge of the species. The key to the resilience of the world. The key to the resiliency and continuity of human society.'[24]

The technological development of artificial intelligence has progressed in recent years from explicit programming to machine learning methods and so-called deep learning (i.e., the use of multi-layer programs such as neural networks to improve machine learning, statistical). The ability to replace people in repetitive, computable tasks is at an all-time high, and it is a significant concern for those in the know to transform the rigid structure of the division of labor by implementing a new Technology Route, rather than easily using technology (machines) to devalue the possibilities for innovation in the labor process and manufacturing experience.

Getting the 'right Technology Route,' in which technology is used to enhance labor skills and productive capacity, requires changing the underlying logic of much of the existing institutional setup around technological research and development and technological innovation, so that technological invention and transformation are monopolized by a few elites and in the process, ordinary people are empowered to participate in innovation on their own. R. M. Unger suggests that limiting new technologies to replacing human beings in repetitive labor and helping to free them to do more work that expands their imagination is the key to reconfiguring the relationship between technology, capital and labor. Only by empowering ordinary individuals to participate autonomously in technological progress based on the knowledge they have learned and

[24] David F. Noble, *Forces of Production: A Social History of Industrial Automation*, New Brunswick and London: Transaction Publishers, 2011, p. 349.

accumulated in their lives and production, it will be possible to move new technologies from 'insular vanguardism' to 'inclusive vanguardism.'[25]

Institutional Innovation and Technical Conditions for Platform Change

However, how can we effectively guarantee the autonomous participation of the working population at prominent and ordinary individuals in the fast-emerging digital, and achieve a change in the platform in favor of shared freedom?

The current development of platforms enabled by digital and intelligent technologies is characterized by centralization, expansion and exclusivity, making it difficult to break the technical character of 'insular vanguardism': firstly, centralization, which is mainly reflected in the ability to make precise decisions from the top down and to create market demand as the capacity to draw resources and research and development increases. This has led to the development of digital technologies that can be increasingly divorced from the real needs of society and reconstruct a new ecology of all aspects of society[26]; secondly, scalability, which is mainly expressed through the expansion of its scale to achieve continuous innovation. In other words, digital technology-based development does not overcome the law of diminishing marginal returns through 'creative destruction,' as Schumpeter put it, but rather through the continuous acceleration of factor inputs to achieve 'a continuous rise in marginal returns'[27]; third, exclusivity. This is the inevitable result of rising returns to scale, so that a small number of technology firms will inevitably become industry giants, and the formation rate of giant scale will be increasingly rapid. The ability of giant platforms to draw resources and create research determines their overwhelming advantage and exclusivity over other market players, such as SMEs.[28]

[25] Roberto M. Unger, *The Knowledge Economy*, London and New York: Verso, 2019.

[26] Lehr, William, and David Clark, Steve Bauer, Arthur Berger and Philipp Richter, Whither the Public Internet?, *Journal of Information Policy*, 9, 2019.

[27] Roberto M. Unger, *The Knowledge Economy*, London and New York: Verso, 2019, p. 29.

[28] R. M. Unger, Isaac Stanley, Madeleine Gabriel and Geoff Mulgan, *Imagination Unleashed: Democratizing the Knowledge Economy*, Published by Nesta, March 2019.

Completely eradicating this developmental logic is onerous, and there are no clear answers. However, ensuring that ordinary individuals have the right and ability to 'voice' in the current development of new technologies should be a direction of institutional design worth pursuing. Here, it is helpful to explore the implications of recent developments in institution building and technological development: the development of empowerment in extensive data tenure and significant data technology solutions that contribute to the autonomy of individuals is a critical institutional innovation to overcome the 'siloed pioneering' nature of new technologies.

As is well known, as new information technology applications become increasingly data-dependent, access to data without human intervention becomes essential. This situation creates two opposing claims in terms of institutional design: the most vital protection for individuals is to grant them private property rights over their data, but this is detrimental to the development of the big data industry, and companies prefer to implement a vague property right that encourages data use and product development. National legislatures are currently struggling to make a clear choice for developing the digital economy—the EU's *General Data Protection Regulation* (GDPR) tends to protect the data rights of individuals, but clearly to the detriment of their industry. In this ambiguous situation, technology platforms with the advantage of technology and capital have been given the resources to expand rapidly. APIs, for example, enable service providers to create digital products (data, software and so on) freely available on the web, and thus, APIs (primarily through the use of cloud computing) help to break down data silos and service compartmentalization, enabling smooth interoperability between platforms, technologies and services. In theory, increased 'interoperability' can help to bridge the gap between different platforms and level the playing field between them.[29] In practice, however, APIs, a digital technology, have become a tool to help giant companies control data and scale. For example, Twitter's API allows developers to build standalone applications that can be directly stripped of Twitter, and Twitter then vertically integrates through mechanisms such as mergers and the creation of application platforms to subsume applications that are better established in the marketplace (e.g., the acquisition of Summize in 2008,

[29] Ian Brown, Interoperability as a Tool for Competition Regulation, OpenForum Academy, November 2020.

Tweetie in 2010 and TweetDeck in 2011 has allowed Twitter to grow in size and market share).

Innovations in data ownership and data technology are necessary to break the monopoly of giant platforms on data and preserve the 'dignity' and autonomy of ordinary individuals in the face of the rapid development of big data technologies. Recent institutional and technological developments have provided the conditions for such a change: (1) in terms of data ownership, many countries have established or recognized the right to data portability to establish the individual as the subject of data rights in Big Data operations (without granting the individual an absolute right to the data). However, this does not give individuals an absolutist private right, to allow room for the development of digital industries, to provide legal support for the free and orderly flow of personal data and to break the monopoly of individual platforms on personal data (e.g., the possibility of using Alipay's Sesame Credit Score to apply for loans from other platforms). In recent years, the European Union, the United States, Singapore, South Korea, India and other countries have introduced laws that provide for the 'right to portability of personal information,' and Article 45 of China's Personal Information Protection Law also comes close to this provision. (2) In terms of data technology, there are new technological innovations that form the premise of supporting the development of big data and respecting the individual's right to data dominant changes. In this regard, 'Solid' (socially connected data), developed in recent years by Tim Berners-Lee, founder of the World Wide Web, deserves continued attention.[30] Solid is a new technological route of Big Data, which is centered on the sharing of information in the context of prioritizing the protection of privacy, in a technological. It is technically possible for users to store their data in freely chosen 'personal online data repositories' (pods); for Solid-approved applications to request data after the user has granted access to the application; and for users to distribute personal information across several pods, both to ensure data security and to prevent a single developer from having exclusive access to the data. Data security prevents a single developer from having exclusive access to

[30] For basic information about Solid, see its official website at https://solid.mit.edu/. For an account of the economic and social significance of the new Technology Route by Chinese scholars, see Jia, Kai, Multiple Technology Routes for a Digital Future, *Culture Vertical*, 6, 2021.

the data. Recently, it has been announced that the UK is using Solid technology to create an 'Open Bank' to promote competition in the banking sector. This new Big Data Technology Route will equip individual citizens with the technological tools to protect their data and increase their autonomy in developing of Big Data.

In sum, in our view, there are two paths to promote the autonomous participation of working people, ordinary individuals, in the process of technological progress, who are complementary and mutually supportive: one is institutional, i.e., to change the 'centralized power without consent' (Shoshana Zuboff) that maintains the monopoly of technological invention and innovation resources by a small elite, so that working people and the wider population can more easily become, or use small and medium-sized market players to share in the resources held by giant platforms. The search for this path of institutional response has just begun. The second is technological, i.e., to create the conditions for a richer Technology Route to be realized and to overcome the centralization, scalability and exclusivity of traditional platform technologies through competition and checks and balances between different Technology Routes, the invention and application of Solid providing inspiration for this route. Without institutional innovation, the legal basis for the monopoly of resources by giant platforms cannot be shaken. In contrast without technological innovation, even if a legal system is established to protect the rights of individual citizens' data, there is little chance of realizing it.

Of course, institutional and policy innovations in a wider range of areas are needed to promote the new Technology Route on broader scale and to realign the relationship between technology, capital and labor. However, this chapter points out that a very crucial step is first to remove the rigid separation between 'intellectual labor' and 'manual labor' from the institutional concept and to effectively safeguard the autonomy of individuals concerning the rapidly developing technology through institutional and technological mechanisms, so as to induce scientific and technological research and development and technological innovation to closely focus on the specific needs of labor and production practices and the accumulation of knowledge. This will ensure that the new technological revolution is genuinely integrated into people's lives and production practices, and will take on new and greater dynamism. Not long ago, the World Bank published the World Development Report 2021, in which it proposed the concept of a 'Social Contract for Data' under the theme of 'Data for a Better Life,' advocating promote trustworthy global systems, regulations and governance frameworks that realize the benefits of data sharing and reuse. This is a good policy initiative to build to stimulate more institutional innovation and technological innovation and to co-create new Technology Routes that serve, rather than devaluing, the autonomy of ordinary people.

This book systematically expands on these understandings. The chapters in the first part focus on the emergence of the economic logic of the platform and its transformative possibilities. The chapters in Part II focus on three levels: firstly, the conceptual discussion of the existence of an ideology that redirects the application of science and technology in China under the governance of the Chinese Communist Party through political ideology, which is a 'double-edged sword' that has the potential to both undermine the constructive function of science and technology and redirect the Technology Route to a direction more conducive to governance goals; then notes that in recent years, in concrete governance practice, over-reliance on top-down 'command-and-control' mechanisms has, in turn, led to an absence of political deliberation in relevant areas, a situation that needs to be urgently changed; and finally suggests that, under the Communist Party of China, the way for China to embrace the development opportunities created by the rise of new information technology and to address the significant governance challenges is to face up to the politics of Technology Route, strengthen the function of political deliberation through institutional innovation and make technology governance subordinate to political deliberation, so as to achieve technology empowerment of the state and promote social openness and cooperation, and to lay the foundation for guaranteeing the 'shared freedom' of different groups of people.

PART I

Rise of Digital Platform and Its Transformative Possibilities

CHAPTER 2

A Brief History of Evolution of Platform-Based Internet

Whether the invention and application of 'technology' is a boon for productivity or a curse for social polarization, environmental pollution, political unrest and many other problems has been a constant source of debate, and the classification of technology seems to be an important way of explaining its differentiated impact. Richard George Lipsey's focus on a particular type of 'General Purpose Technologies,' for example, explains the important role of technology as a fundamental driver of long-term economic growth; Lewis Mumford's typology of 'authoritarian technologies' and 'democratic technologies' reveals the political content of technology while also alluding to its different impacts on human society. Daron Acemoglu's distinction between 'Enabling Technology' and 'Replacing Technology' explains the 'decline' in wages for all after the 1980s. The 'Great Reversal,' in which wages for all began to fall after the 1980s, is difficult to explain within the framework of Jan Tinbergen's traditional analysis of 'skilled workers' and 'unskilled workers.' This phenomenon is difficult to justify within the framework of Jan Tinbergen's traditional analysis of 'skilled' and 'unskilled' workers.

In their theoretical vision, 'technology' is not a single concept but a composite concept that includes different types of objects. It is based on the distinction between types of technology that embody different attributes or differentiated influences that we can consciously choose

Y. Jiang and K. Jia, *Transformation of Platform Governance in China*, IPP Studies in the Frontiers of China's Public Policy,
https://doi.org/10.1007/978-981-99-6456-7_2

a particular Technology Route to adapt it to the aims and intentions of decision-makers. In this sense, technology is not limited to a value-neutral objective phenomenon but evolves into a 'man-made object' that embodies a particular value orientation.

The consensus on the current development of the Internet (or more generally the 'cyberspace') is that we have now developed a digital business structure with digital platforms at its core. This is even though in the early days of the Internet, distributed, loosely organized online communities were the dominant form. The emergence of global collaborative phenomena such as open-source software and Wikipedia, as well as new businesses such as end-to-end services (e.g., Napster), portal news sites and social media, all primarily embodied community characteristics and distributed innovation patterns and have been sustained. However, with the rise of digital business platforms and the formation of monopolistic structures, the 'digital platform structure,' centering on the digital platform to control core bottleneck resources while opening up peripheral resources to other ecological players, has become the new dominant structure.

The political impact of the two digital ecological structures, online communities and digital business platforms, is very different, with the former relying more on the spontaneous behavior of decentralized subjects. At the same time, the latter is more controlled and organized. It is not the intention of this chapter to compare the differences between these two structures in terms of their impact on the overall level of social welfare or the welfare of different groups—although this is also important—the more prerequisite question is how we actually move from the structure of online communities to the structure of digital business platforms and how this process takes place. How did this process take place? In particular, concerning the subject of this section, what role has the 'technological' evolution of the Internet itself played in this process?

This chapter will focus on analyzing of three landmark technological issues around this question. By tracing the historical process of technological evolution, the first section attempts to reveal how the logic of the Internet's technological development will drive the transformation of online community structures into digital platform structures. It is worth noting that while many discussions tend to view 'platform' as a change in organizational structure, in the Internet context, 'platform' itself implies the choice and design of technological architecture, where technological evolution and organizational evolution form the evolution of technology and the evolution of the organization forms an inseparable unified whole. In this process, we can understand that the seemingly objective and neutral evolution of technology in fact still reflects the richness and far-reaching influence of technopolitics.

The Prologue to the Digital Platform Unfolds in the Decline of the IBM PC

The evolution from mainframes to personal computers did represent not only a difference in the form of computers but also a change in the ecological structure of the two industries. It was the formation and development of the personal computer that led to the formation of the 'digital platform' as the core of the business structure. It really started the historical process of the digital ecology flourishing.

IBM's integrated design and manufacturing capabilities, covering hardware and software modules, shaped the S/360's success and its position as the 'Blue Giant' in the computer market. However, IBM's glory did not stop there, and the IBM Personal Computer was launched in 1981, pushing it even further to the top. However, the value of the IBM Personal Computer was not limited to IBM itself. It was perhaps even more important for Microsoft and Intel, as it finally ushered in a new era of 'digital platforms.'

The IBM Personal Computer was not the first; the first West Coast Computer Faire opened in San Francisco in April 1977, and it was Steve Jobs' Apple II and Commodore's PET personal computers that are widely regarded as the beginning of the personal computer and widely regarded as the starting point of the personal computer. Despite the buzz, controversy or doubt still surrounds the question of whether the PC is the future, and the decision of IBM CEO Frank T. Cary ultimately set the stage for the subsequent glory of the PC market. Unlike the centralized and time-consuming nature of mainframe development, the rapid pace of renewal in the PC market placed entirely different demands on the computer production process and management organization. Even at the very beginning, the pace of product renewal in personal computers was swift. After the Apple II and PET debut in April 1977, Tandy introduced the TRS-80 and the first video game in a personal computer in August.[1] It was in this context that Microsoft and Intel boarded the IBM Personal Computer ship, driven by Carey and William Lowe, who proposed a new path for the product development process unlike any other in IBM's history, namely to bypass the tedious internal section process and invite a wide range of partners to join in an open architecture. If the IBM S/360 mainframe gave birth to the classic book The Mythical Man-Month: Essays on Software Engineering, the IBM PC created 'co-engineering'. Both represent a new exploration and core experience of the division of labor between different actors in the digital production process.

It is worth noting that in the late 1970s, neither Microsoft nor Intel was the strongest players in their fields. For Microsoft, Digital Research was the market leader in personal computer software development at the time; for Intel, Motorola and Zilog could produce better integrated circuits.[2] But once again, coincidence and uncertainty played a significant role in a significant historical event. Although Lowe and the others first approached Microsoft, the latter did not convince the former and so turned to Digital Research, with whom they again failed to reach a consensus. The core of Lowe's brief was to acquire the operating system needed for PCs as a service for a limited period. However, neither Microsoft nor Digital Research thought that the considerable investment in an 'operating system' would yield sufficient returns. Fortunately, Microsoft changed its mind and fulfilled IBM's requirements by acquiring another company that happened to have a specific base, which was the precursor to the 'DOS' system.

With the vast success of IBM's personal computers, Microsoft and Intel have officially entered the history books. While IBM remains the 'definer' of personal computer products, Microsoft and Intel have become increasingly influential in the open ecosystem. It is worth noting that IBM's contract with Microsoft gave Microsoft significant openness in that IBM did not restrict Microsoft from selling the operating systems it developed for IBM PCs to other PC manufacturers, probably in large part due to the antitrust investigation being conducted by the US Department of Justice

[1] The Consumer Electronics Hall of Fame: Tandy/RadioShack Trs-80, https://spectrum.ieee.org/the-consumer-electronics-hall-of-fame-tandyradioshack-trs80-model-1.

[2] Windows + Intel = Wintel?, https://www.mayin.org/ajayshah/MEDIA/1998/wintel.html.

against IBM at the time.[3] Interestingly, after the Reagan administration dropped its case against IBM in 1982, IBM moved away from an open strategy and developed all its integrated systems. In January 1983, Time magazine's Person of the Year featured 'The Personal Computer' on its cover for the first time.

But for IBM, the personal computer market that it had kicked off did not necessarily mean he would always be a leader. IBM's traditional corporate organization and product-oriented production philosophy were not adapted to the fast-moving personal computer market. After the success of the initial generation of products, IBM reintegrated the personal computer division into the company's section organization in January 1983. However, the slow decision-making process began to affect the agility of the personal computer division in responding to the market. The more severe problem was that IBM management did not realize the essential difference between the personal computer and other products such as mainframes.[4] In their view, PCs and their applications should follow a quality control process similar to that of other products and be brought to market without 'bugs'; on the contrary, the emphasis was on getting to market faster and having users identify problems as they were applied. This is the secret of the success of open-source software. The difference in product development patterns between mainframes and PCs, in fact, reflects a fundamental difference in the two production models: the former emphasizes centralized development, while the latter relies more on open architecture. Only with an open architecture can the PC attract enough software developers and participants to enrich the scenarios and functions of the PC, which is an entirely different 'ecological' structure from the mainframe.

The slowness or conservatism of the production process and organizational model is only one symptom of IBM's strategic mistake. Another critical issue is the neglect of 'compatibility' in the design of its products. Unlike traditional products, the value of a personal computer depends to a large extent on the wealth of application software, and this dictates that the producer of a personal computer needs to form a symbiotic digital ecosystem with the developers of application software, who need to take

[3] The Rise of the Wintel Empire, https://abcnews.go.com/Technology/story?id=98348&page=1.

[4] How the IBM PC Won, Then Lost, the Personal Computer Market, https://spectrum.ieee.org/how-the-ibm-pc-won-then-lost-the-personal-computer-market.

into account the cost of porting the latter when launching new products 'went the other way'. The incompatibility caused significant migration costs for the other ecological players while also causing them to reflect on their 'right to define' the market for IBM's personal computer products. Previously, IBM had often been seen as the definer of PC products, with other manufacturers following suit and thus achieving economies of scale in the PC market. However, PC Junior's incompatibility eventually pushed the market into the arms of Microsoft and Intel, who joined forces (i.e., the Wintel Alliance) to provide a 'new standard' for the PC market. The combination of the latter two (i.e., the Wintel Alliance), while providing a 'new standard' for the PC market, also ultimately established the dominance of the open platform architecture. Unlike IBM's product thinking, the Wintel Alliance provided the PC market with a standard 'chip+ operating system' platform on which PC manufacturers defined their products and software developers developed their software products or digital services based on compatible interfaces without the latter having to worry about the 'unpredictability' of the PC products produced by the former. The latter do not need to worry about the cost of dedicated investment due to the 'incompatibility' of the PC products produced by the former. The new developments in the PC market, based on the Wintel Alliance, have genuinely opened up the dominance of the 'platform' architecture in the digital world, and this has gradually extended to all areas of 'cyberspace'.

The Rise of the Mobile Internet and the Evolution of Digital Platforms

Along with the spread of smartphones, the PC-based Internet is rapidly changing into the mobile Internet. This process does mean not only a change in how people are connected, but also an increasing tendency for the evolution of technology to become platform-based, resulting in the emergence of mega-platforms. In terms of market capitalization, many digital platform companies are already larger than the gross national product of most countries. In December 2021, for example, Apple's market capitalization is already close to \$3 trillion, surpassing even the total GDP of the UK in 2020. Remarkably, this change process has only taken place over a mere 15 years—Apple's market capitalization prior to 2006 was even less than \$10 billion. On the face of it, it is the burgeoning digital business service and the vast number of users who use it that underpin the considerable market capitalization of digital platform companies. However, the platform-based transformation of digital technology is an equally important factor that cannot be ignored.

In 1996, John Perry Barlow, an early evangelist for the Internet and founder of the Electronic Frontier Foundation, published a Declaration of the Independence of Cyberspace, which expressed most eloquently the idea that the Internet transcends geographical borders and is therefore exempt from the jurisdiction of sovereign states. The political idea is that the Internet transcends geographical borders and is therefore exempt from the jurisdiction of sovereign states. 'Where we gather, you have no sovereignty'.[5] This succinct statement not only expresses the 'utopian' aspirations of cyber-liberals, but also implies the formation of a new globalized cyberspace. The fundamental reason for Barlow's idealistic expectations of the Internet lies in the technical aspect—the open 'peer-to-peer' architecture of the Internet, which allows anyone to access cyberspace without restrictions or limitations. This also formally bypasses the influence and control of traditional 'sovereign' authority based on geographical boundaries. Indeed, it is based on this 'end-to-end' open architecture, and new 'shared' businesses such as Napster have emerged. Despite the controversies and questions about piracy, the end-to-end structure has functionally enabled the direct connection of information or the direct exchange of resources between users. The 'blockchain' era, which emerged in 2009, can also be seen as a newer version of the 'end-to-end' structure.

However, the 'end-to-end' structure is not the end of the development of the Internet. The 'platform' structure that emerged during the production of personal computers has gradually been transplanted to other technological fields to the point where it has become platform-based. Compared to the 'end-to-end' architecture, the 'platform' architecture inherits its characteristics of openness, interactivity and extensibility; at the same time, however, the critical difference of the 'platform' architecture is that it makes full use of digital technology. The critical difference, however, is that it takes advantage of the 'systemic' nature of digital technology and controls the core or critical layers to influence the other layers.

[5] John Perry Barlow, A Declaration of the Independence of Cyberspace, https://www.eff.org/cyberspace-independence.

In the PC Internet era, the 'systematization' of digital technology is characterized by two dimensions: technology hierarchy and networking. On the one hand, the technology hierarchy refers to the fact that the digital technology system is composed of a series of layers, with the upper layer often relying on the lower layer, so that the lower layer can be regarded as the infrastructure of the upper layer. In compassion, upper layer is based on the particular applications of the lower layer. In the case of the PC Internet, the bottom-up technology hierarchy can be roughly divided into hardware, logical, middleware, application and content. The hardware layer includes computing chips, storage devices, interface buses and so on; the logical layer includes signal transmission protocols, operating systems and so on; the middleware layer includes computing frameworks, application interfaces and so on; the application layer includes system native applications, browsers and so on; and the content layer specifically covers media, information and other industries. In the PC Internet era, the Wintel alliance seized the two critical links of computing chips and operating systems at the hardware and logic layers, respectively. It achieved the scale effect through the 'open up' platform strategy, which also enabled it to maintain its control over the PC industry. On the other hand, the 'systemic' character of digital technology is also reflected in the extension from personal computers to networking. The PC as a 'product' stands alone and covers the five technical levels mentioned above or 'Software as a Service (SaaS),' 'Backend as a Service (BaaS),' 'Infrastructure as a Service (IaaS)' and other series of cloud computing architectures. The networked system covers the same five layers mentioned above, except that it may present different functional requirements and consequently reflect differentiated technical forms. For example, at the operating system level, personal computers are dominated by Microsoft's Windows operating system (Windows), but it is not adapted to the functional requirements of servers, which are more often met by 'Linux' or 'Apache Tomcat.'

Compared to the PC Internet, the mobile Internet extends and enriches the two dimensions of technology level or networking, and develops the third dimension of 'application services'. In the technology-level dimension, the typical change in the mobile Internet is the rise of the Android and iOS operating systems, which not only impacted the monopoly of the Windows operating system, but also reshaped the inter-relationship between the different technology levels. In particular, the evolution of the Android operating system as open-source software has considerably freed up the competitive space for smartphone manufacturers, giving them greater bargaining power in the game of hardware manufacturing versus operating systems. In the networked dimension, the mobile Internet has brought about a richer range of product systems, extending from personal computers and smartphones to wearable devices, the Internet of Things and many other types. It has also accelerated the digital transformation process. In addition to these two dimensions, the new mobile Internet change is that 'application services' are gaining importance as a third dimension. In the PC Internet era, different application services were not strongly related to each other, with search engines, e-commerce, social media and other application services being independent of each other; after entering the mobile Internet, it has become the norm for different application services to be nested with each other, and multiple layers have been created in the process. For example, application services such as mobile application shops, mobile payment and maps have become the new 'infrastructure' to support the development of other application services. At the same time, vertical applications such as instant messaging and taxi services can also absorb other application services to build a digital ecology by their massive user base. In this context, cross-border applications and competition have become the norm in the mobile Internet era. A market study found that most mobile app installers have expanded by tens or even hundreds of times in the decade between 2010 and 2022.[6] For example, the WeChat installer has expanded 575 times from version 1.0 in 2011 to version 8.0.24 in 2022, the QQ installer has increased 37 times in eight years, and the Jitterbit installer has grown six times from its introduction in 2019 to 2022. A closer analysis of the specific contents of the installer reveals that the vast majority of files and code are used to implement cross-border functionality. For example, WeChat as an instant messaging app is gradually adding payment, e-commerce and information functions, ShakeYin as a short video is gradually adding group chat and e-commerce functions, and Drip as a taxi app is also adding e-commerce and payment functions, all of them are constructing and reshaping the hierarchical and systematic structure of application service dimensions.

In contrast to the 'end-to-end' structure, digital platform's key is controlling the 'bottleneck' resources of the digital technology system

[6] Inflated by 575 times in 11 years, why are mobile app installers getting bigger?, https://mp.weixin.qq.com/s/1XBhcnDC0Ahm-QYzmNqYxA.

to leverage and define the value creation behavior of other ecosystem players. In the PC era, the 'bottleneck' resources were only the central processor chip and the operating system, making the Wintel Alliance the core platform and defining the logic of the evolution of PC products and the digital business ecosystem they were built. In the mobile Internet era, the 'bottleneck' resources have gradually moved up the technology hierarchy. They are widely manifested in multiple forms such as browsers, mobile application shops, mobile payments, maps and super apps (e.g., WeChat). In the process of fighting for control of these resources, a variety of digital platforms have emerged from different areas. In the competition for control of these resources, various digital platforms have emerged from different areas.

Based on an analysis of the evolution of the PC Internet to the mobile Internet, it is easy to see that although the Internet boom started with the innovation of 'end-to-end' technology architectures, this is far from the end of the evolution of digital technology. The emergence and dominance of platform-based technology architectures largely reflect the competition between multiple Technology Routes, and the attributes of technology politics are evident in this process. However, even so, two complementary points are crucial.

On the one hand, although the dominance of platform-based technology architectures is a reality, we still cannot consider it an inevitable or inevitable technological trend. As we have seen in the previous section on the reflection and competition between the World Wide Web and Solid, the rise of platform technology architectures can be largely attributed part to the particular design of the World Wide Web that integrates applications with user data. This is not the only solution, as Timothy John Berners-Lee's reflections on the 'data shaft' phenomenon show. The Solid solution, which separates the two, explores an 'alternative path'. It is reasonable to imagine that if Solid had been Berners-Lee's original solution in 1990, rather than the World Wide Web, the rise of platform-based technology architectures might not have been as 'natural' as it seems—and this once again illustrates the multiplicity and politics of the evolutionary path of technology itself.

On the other hand, even if platform-based technological architectures become dominant, this does not mean there is no room for online communities based on 'end-to-end' architectures. While open-source software and Wikipedia seem to be the only representative examples, and digital platforms such as Uber and Airbnb, with their 'sharing economy' trappings, seem to predetermine the construction of communities with a common ethos as a 'utopian' fantasy, these are not enough to make a case for an 'alternative path'. However, these are not enough to pronounce a death sentence on the 'alternative path'. Indeed, Bitcoin in 2009 and the subsequent boom in blockchain-related businesses still represent a powerful search for freedom of organization from central control. More fundamentally, if we look beyond the superficial veneer of platform architecture to a handful of glamorous mega-platform companies and take a broader view of the digital transformation of industry, it is easy to see that the ubiquitous yet vibrant open-source software and open knowledge underpin the digital transformation of all scenarios and sectors—and this is exactly what Google's chief economic officer has done. This is what Google's chief economist Hal Varian once called 'Varian's Law'. According to Varian and others, the exponential growth of technological innovation in the digital age is not due to the uninterrupted discovery of 'disruptive innovations,' but to the combination of innovations based on vast amounts of open-source software.[7] Perhaps the production and dissemination of knowledge based on an 'end-to-end' architecture are the huge 'iceberg' that lies beneath the 'sea level' of digital platforms.

It is on the basis of these two reflective analyses: we need to realize that digital platforms are also not the 'end game' of the Internet; otherwise, we would only see competition and monopoly over key bottlenecks and no space or value for mutually beneficial or altruistic cooperation. However, such limited thinking, in fact, pushes platform-based architectures into a hopeless future, a dilemma reflected in the widespread global criticism of mega-platform companies over the past two years. Against this backdrop, the exploration of a new Technology Route is not only important but urgent.

[7] Shapiro, Karl, Hal Van Rijn, *Information Rules: A Strategic Guide to the Network Economy*, translated by Meng, Zhaoli and Niu, Luqing, Peking: Renmin University of China Press, 2013: 4-5.

CHAPTER 3

Analysis of Platform Economic Logic: Structure and Its Future

Uber, Airbnb and Salesforce are still representatives of the power of Silicon Valley. However, the rise of Chinese companies such as Lufax, Drip, Youku and Meituan has made this model of industrial organization rapidly popular in China. The business models of many companies, including Google, Facebook, Amazon, Baidu and Taobao, can be included in the scope of the 'platform economy,' but it was not obvious in the PC era and only after entering the mobile era did it show its impact on the traditional industrial organization model, as well as on the creation of value and the development of a new industry. Only after the mobile era did the impact on the traditional industrial organization model and the reconfiguration of the value creation and distribution model become apparent.

However, the question is, what kind of industrial organization model can be called a 'platform economy'? Furthermore, how does it differ fundamentally from classical theory in terms of economic theory? This is the main question that this short article seeks to answer. It is important to note that the impact of the rise of the 'platform economy' is not limited to the field of industrial organization, but has equally far-reaching implications for social policy, tax policy, competition and antitrust policy.

The 'platform economy' banner many long-standing theoretical frameworks and policy practices are in danger of becoming or are facing failure

Y. Jiang and K. Jia, *Transformation of Platform Governance in China*, IPP Studies in the Frontiers of China's Public Policy, https://doi.org/10.1007/978-981-99-6456-7_3

and challenges. For this reason, it is of great importance to clarify the theoretical meaning of the 'platform economy' and understand its operation's basic logic to reshape the regulatory system to meet the new round of Industrial Revolution.[1]

The Difference Between Platform Economics and Traditional Economics

What exactly is a 'platform'? In fact, the answer to this question is controversial. For example, the Apple App Store and Youbou are both considered 'platforms,' but there is a very clear difference between them: in the Apple App Store, developers are free to set prices for their apps, whereas on Youbou drivers have to accept pre-determined prices from the system. For example, while the cars on Shouqi are owned by the company (i.e., the seller), the cars on Drip (and Youbou) are all private cars. The question that naturally arises is whether they can be called 'platforms'? If not, how do the differences affect the logic of the 'platform economy'?

In fact, 'platform' has become more of a marketing language, and many of the models of industrial organization that bear the name 'platform' are not strictly economic concepts. In his seminal paper with his co-authors, Tirole summarized three elements of a 'platform economy': the existence of two or more sets of customers (e.g., drivers and passengers), the existence of network effects between different customers (e.g., the more drivers on a platform, the easier it is for passengers to get a ride on that platform) and the need for intermediaries (i.e., the need for a platform to match drivers with passengers). David Evans, a professor at the University of Chicago Law School, and one of the first scholars to study the economics of platforms, summarizes the meaning of 'platform' similarly.

The three elements mentioned above may theoretically clarify the meaning of the 'platform economy,' but they do not yet clarify the key elements of the 'platform economy' in an industrial organization model. In other words, what role does the 'platform' play in trading? Farmers'

[1] The following is mainly summarized from the following literature and discussed in detail in Chapter 4 of this book: Jean-Charles Rochet and Jean Tirole, Platform Competition in Two-sided Markets, *Journal of the European Economic Association*, 1(4), June, 2003, 990–1029. David Evans and Richard Schmalensee, *Matchmakers: The New Economics of Multisided Platforms*, New York: Harvard Business School Press, 2016.

markets can also play a role in brokering transactions, so can they, therefore, be called 'platforms'? Tirole concludes his article by explaining that the key to an organizational model being called a 'platform' is the 'non-neutrality' of transactions. In other words, the 'platform' exists because it can effectively employ a cross-subsidization strategy and can impose different pricing strategies on sellers and buyers and influence them differently. In other words, a 'platform' exists because it can effectively employ cross-subsidization strategies and impose and influence different price strategies on sellers and buyers. A market that satisfies this condition is called a 'bilateral (or multilateral) market,' as opposed to the 'unilateral market' of traditional classical economics, and this leads to the main difference between 'platform economics' and traditional classical economics. The main difference between 'platform economics' and traditional classical economics is the question of price structure.

The Key to the Platform Economy: The Price Structure

Under traditional economic theory, the equilibrium price in a perfectly competitive market is equal to marginal cost. In contrast for an imperfectly competitive oligopolistic market, the equilibrium price satisfies the Lerner equation, which states that the difference between price and cost is inversely proportional to the elasticity of demand. For platform economics, however, one of its main conclusions is that **the pricing of profit-maximizing products is not directly related to marginal cost but rather to price structure rather than the price level**. Thus, platform economics parted ways with traditional economics.

Take the example of an app store (app store). Developers upload apps to the platform, and users buy them on the platform. The app store can charge either the developer (the seller) or the user (the buyer), the sum of which constitutes the 'price level' of the platform, and the relative ratio of the two is referred to as the 'price structure.' The analysis by Tirole et al. shows that the price charged to either side (buyer or seller) in a profit-maximizing platform does not depend directly on variable costs (and thus differs from conventional economics) but is proportional to the bilateral elasticity of demand. This leads us to an important characteristic of platform economies (and bilateral markets): **indirect network effects** (sometimes called 'cross-network externalities').

Indirect network effects are those where the size of the network on the one side (e.g., the buyer) will impact the size of demand on the other side (e.g., the seller). For example, the more apps there are in an app shop (seller size), the greater the demand from users; conversely, the greater the number of users of an app shop (buyer size), the more programmers will be willing to develop apps on that app shop platform. However, it is important to note that the size of the indirect network effect (i.e., the cross-demand elasticity) may differ for the bilateral parties (buyers or sellers). In other words, the 'seller's influence on the buyer' may be different from the 'buyer's influence on the seller,' and because of this, the price charged by the platform to both parties may differ. Fig. 3.1 shows the possible equilibrium price states on a monopolistic platform.

The two lines in each subplot in Fig. 3.1 are the change in the platform's optimal pricing for one side (e.g., the seller) relative to the price for a given other side (e.g., the buyer), the intersection of which then reflects the platform's optimal price structure for a given level of externalities. Figure 1A indicates that the indirect network effects are the same for both sides and prices are symmetric; in Figure 1B, the demand influence of side 2 (e.g., the buyer) on side 1 (e.g., the seller) increases, and the platform will reduce pricing on side 2 to attract (or maintain) more buyers to join the platform while increasing pricing on side 1 to achieve optimal profits on the seller; further, the demand influence of side 2 on side 1 increases further, the influence of side 2 on the demand of side 1 is further increased, so that it is profitable for the platform to adopt even negative pricing (e.g., subsidies) for side 2, as this will then further stimulate the demand of side 1 and thus generate revenue. Take the example of online taxi platforms. Passengers have multiple options to travel, so the elasticity of demand for online cars is less than the elasticity of demand for passengers, so platforms tend to heavily subsidize passengers to attract buyers to join the platform (which means that the platform imposes 'negative pricing' on passengers) while imposing high prices on online cars. On the other hand, as user habits develop, the elasticity of passenger demand for online taxis gradually becomes greater so platforms can gradually remove subsidies (which means that they gradually raise prices for passengers) to make more profit.

More complex analyses (e.g., the multiple ownership of buyers or sellers, the ownership of platforms and so on) are not overly covered in this short review, but it is the difference between the 'platform economy' (and platform economics) and traditional classical economic theory

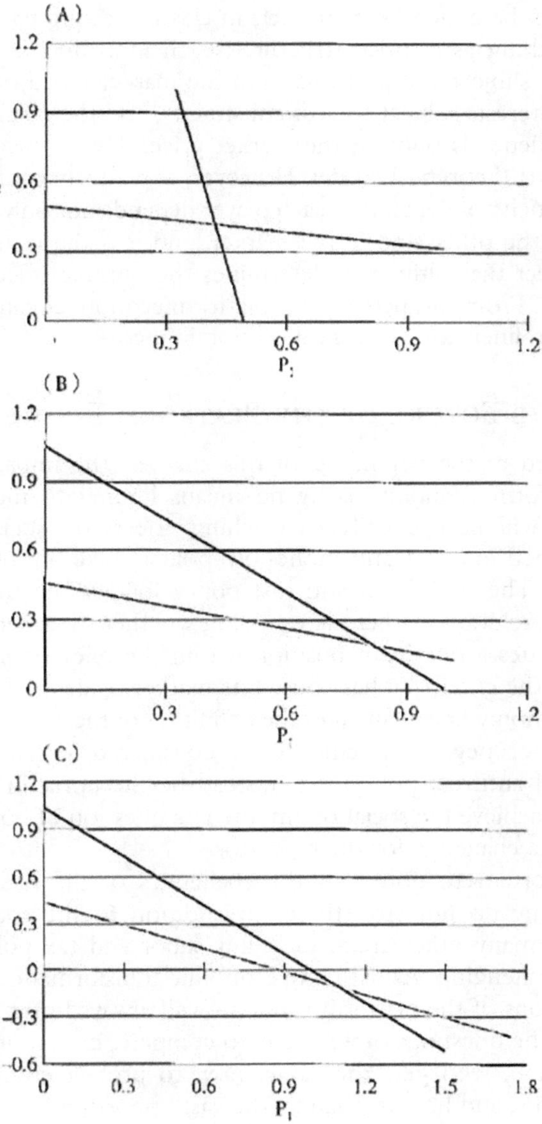

Fig. 3.1 Platform pricing diagram

that needs to be added here. In fact, in classical economic theory, there is no such thing as a 'price structure'; each individual is treated as an independent subject and participates in the market based on maximizing individual interests subject to cost constraints, and the resulting elasticity of demand depends only on the market price. This is an oversimplified and unrealistic theoretical model. However, as platform economics points out, the elasticity of demand of each player depends not only on its market but also on the other side of the market, and it is the resulting indirect network effect that ultimately determines the optimal price structure of the platform. From this perspective, platform economics can even be seen as a further refinement of classical economic theory.

The Future of the Platform Economy

As mentioned at the beginning of this chapter, the impact of the rise of the 'platform economy' is by no means limited to industrial organization; it will have equally far-reaching effects on social policy, tax policy, competition and anti-monopoly policy. Take antitrust policy as an example. The traditional antitrust policy follows the three stages of 'defining the relevant market, determining whether the respondent enterprise constitutes a dominant position within the relevant market and, if so, whether the enterprise has abused its market control.' The rise of the platform economy first challenges the definition of the relevant market. At the same time, 'negative pricing,' often considered 'unfair competition' in traditional antitrust policy, may instead be the norm in the platform economy to achieve the social optimum. The question of how to redefine this remains a challenge for the regulator.

It is important to note that the challenges of the rise of the platform economy do not stop there. In addition to the specialized area of antitrust, many other areas, including labor and tax policies, will be extremely challenging. As with any economic transformation, the disruptive innovations of the platform economy will always bring both winners and losers. The questions of who gets to compete, how to innovate, how to create value, who gets the value, how to protect communities and workers' rights, and how to channel the vast opportunities for innovation that arise in the course of dramatic change will then become increasingly acute and unavoidable. Indeed, there is no such thing as a market without rules, and the winners and losers are often determined by who gets to compete in which way. Given this, how will the power structure change when the platform owners set the very important rules of the game in an uncontested manner? In other words, in the platform economy, the power to make rules may be shifting from governments to platforms and the algorithms that underpin them, and all the political struggles surrounding the making of these rules will be part of what defines markets and society. Some of the struggles will be over the data protection, consumer and worker rights, and others will be over how to regulate business models that engage in institutional arbitrage. In these cases, what should we make of the emerging disruptive innovations? Are they creating diverse entrepreneurship, or are they dismantling social security systems? Can they create new sources of income and provide fair compensation for the displaced? Are policies that protect workers and encourage innovation sufficient to provide a dignified and sustainable standard of living for citizens? In the age of the platform economy, will there be another Henry Ford declaring that everyone should receive enough income to buy his or her products? The answers to none of these questions will be too simple, and the economic restructuring around platforms will inevitably reconfigure the landscape of interests and ultimately influence the development and implementation of laws and codes. In short, as Jose van Djick points out, these struggles will intertwine and reshape our communities and social life, and they will also reshape our markets and competitive structures.

CHAPTER 4

Rethinking Platform-Based Internet: Possibilities of New Technology Routes

Rethinking the 'Internet' refers to a reflection on the current state of technology in the Internet, or 'cyberspace,' which includes both 'speculation' on other technological solutions that have not historically emerged or been mainstreamed and 'speculation' on the multiple Technology Routes that may emerge in the future development of the Internet. This includes 'speculation' about other technological solutions that have not historically emerged or been mainstreamed and 'speculation' about the multiple Technology Routes that may emerge in the future development of the Internet. Both begin with a dissatisfaction with the current technological solutions or development models of the Internet and an attempt to build on them to explore 'better' technologies. On the other hand, it is, in fact, difficult to define 'better technology,' so rather than exploring multiple possibilities in order to find 'better technology,' the idea is to present the options for 'better technology' in order to make it possible to find 'better technology,' so that different subjects can compare and choose within a larger space of intellectual emancipation. In this sense, this chapter will analyze the new Technology Route from two perspectives.

The range of technologies, from blockchain to Web 3.0, focuses on specific types of technology. In fact, from the end-to-end architecture of the early days of the Internet to the platform-based architecture, to blockchain and Web 3.0, we have gone through a 'recursive cycle'. While

Y. Jiang and K. Jia, *Transformation of Platform Governance in China*, IPP Studies in the Frontiers of China's Public Policy, https://doi.org/10.1007/978-981-99-6456-7_4

inheriting the 'end-to-end' architecture, blockchain and Web 3.0 have incorporated and recombined new technological features to achieve more radical distributed innovation, which offers new possibilities to break the digital platform dilemma. On the other hand, unlike the analysis of specific types of technology, the 'fork' discussion is not concerned with the new technology itself but with the important 'methodology' for finding and implementing it. Based on the evolutionary realities of open-source software, the fork perfectly illustrates the birth of a new Technology Route in the digital age. The discussion around 'forks' will provide a compass for exploring the new Technology Route.

From Blockchain to Web 3.0

Since the birth of Bitcoin in 2009, blockchain and its derivative technologies have been on a 'tear'. The reason why blockchain has attracted such an influx of capital and talent, resulting in such a large and rich body of technology, can essentially be attributed to people's expectations and enthusiasm for the transformative power of distributed technology and the imagination of the industry—and it is in this regard that we can see blockchain as 'recursive reincarnation' of the 'end-to-end' architecture.

As blockchain's intellectual and technical origin, Bitcoin is a perfect example of the value and technical logic of distributed architecture. Bitcoin was born out of a long-standing attempt by the 'Cypherpunk' network community to implement a cryptocurrency. On 31 October 2008, 'Satoshi Nakamoto'[1] published a paper in the Cypherpunk community entitled 'Bitcoin: A Peer-to-Peer Electronic Cash System,' and in January of the following year released the source code for the Bitcoin client, which contains approximately 31,000 lines, on SourceForge.net,[2] thus marking the technical birth of Bitcoin. This marked the technical birth of Bitcoin.

[1] So far, the true identity of 'Satoshi Nakamoto' has not been confirmed. Some even believe that Satoshi Nakamoto may not be a single person but a group of people. However, either way, the founder of Bitcoin remains a mystery to this day. However, identity of Satoshi Nakamoto is less important than the 'anonymity' sought by the 'cypherpunk' community, not least because of the need for privacy. It is not only a requirement for privacy, but also an insistence that 'anyone can be Satoshi Nakamoto.'

[2] SourceForge.net is a mainstream open-source software website where the source code of many open-source software is uploaded for interested programmers to read and develop.

In terms of evolution, the crypto-punk community was born out of small gatherings organized by Eric Hughes,[3] Timothy May[4] and John Gilmore[5] in the San Francisco area in the 1990s, sparked by opposition to the US government's post-World War II suppression of cryptographic research and application. At one of these gatherings, the writer Jude Milhon first coined the term 'Cypherpunk,' derived from the terms 'Cipher' and 'Cyberpunk'. 'Cipher' and 'Cyberpunk' were combined. As participation grew, Gilmore and others started a mailing list in 1992. 1992 saw the publication of Eric Hughes' 'The Cypherpunk Manifesto,'[6] which focused on the ideas and values of the group and led to the formation of a more stable crypto-punk community. In one sentence, the crypto-punks' aspirations can be summed up by their commitment to building an anonymous society.[7] In their view, privacy is the foundation of civil rights in the information age. It is only through an open society based on privacy protection that social relations can be reconstructed. The nature of the individual can be released, and humanity can truly enter a free society in the information age.

Unlike any previous era, the development of digital technology has led the crypto-punk community to believe that it is possible to innovate with code to truly implement an anonymous society's vision. Given the importance of 'Nationalized money' in modern society, achieving an electronic currency is both 'decentralized' and 'anonymous,' bypassing central banks and banks as the central medium. The crypto-punk community has been working on this issue for a long time. For a long time, however, the development and design of electronic currencies have struggled to combine these two goals—until the creation of Bitcoin. Technologically and logically, Bitcoin's core innovation is a 'combination' of traditional cryptography, 'end-to-end' network connectivity and communication, a distributed database and a proof-of-work mechanism, which allows for the issuance and transaction of currency without relying on

[3] Professor, Department of Mathematics, University of California, Berkeley.

[4] Senior Scientist, Intel.

[5] Co-founder of the Electronic Frontier Foundation.

[6] Eric Hughes, A Cypherpunk's Manifesto, http://www.activism.net/cypherpunk/manifesto.html.

[7] Eric Hughes, A Cypherpunk's Manifesto, http://www.activism.net/cypherpunk/manifesto.html.

a third-party entity. The core innovation is implementing a 'combination of traditional cryptography, end-to-end' network connectivity and communication, distributed database and proof-of-work mechanism.

The technical success of Bitcoin has unleashed the imagination of distributed technology architectures, all of which have been distilled into the technical concept of 'blockchain' and have led to a series of ongoing innovations such as Ether, Non-Fungible Token (NFT) and Web 3.0. The first of these is the Ether that is the successor to Bitcoin. Firstly, while inheriting the basic framework of Bitcoin, Ether replaces the 'numbers' in the blockchain data structure, which represent the number of currencies, with 'contracts,' thus greatly expanding the application space of the blockchain and making it a 'general-purpose technology'. The basic characteristics of a 'General Purpose Technology' are wide range and many kinds of applications. The views of Vitalik Buterin, the founder of Ether, most clearly demonstrate this potential. For Buterin, 'a blockchain is a platform consisting of a set of protocols to which anyone can upload a program and have it run automatically, with the historical state of the program being recorded and made publicly visible, and the platform ensuring that the program will persist according to the platform's established rules, following crypto mics'.[8]

Secondly, compared to the traditional thinking of Bitcoin and Ethereum, which see blockchain as a network for storing and disseminating 'undifferentiated data,' the breakthrough of NFT is that it uses blockchain technology to define, identify and trade 'differentiated data.' For a long time, the basic logic of digital transformation has been to correspond the real space to the cyberspace, in which the complexity of reality is homogenized into undifferentiated 0/1 data, and to improve the operational efficiency of the whole society through the processing and analysis of the data; however, it is obvious to us that the whole society is not all homogenized objects, even for the same goods, the differentiated emotions given to them by different people will make them 'unique'. The digital transformation does not reflect this 'incommensurability,' and it is here that NFT has made a technological breakthrough based on blockchain. This is where NFT achieves a technological breakthrough based on blockchain. The so-called incommensurable token is, in fact, a data traceability process based on the blockchain, which allows any subject

[8] Buterin, V. Visions part I: The value of blockchain technology, https://blog.ethereum.org/2015/04/13/visions-part-1-the-value-of-blockchain-technology/.

in the network to confirm ownership of the data through a smart contract without restricting the opening and use of the data. In this sense, NFT technically realizes the separation of the 'bundle of rights'; that is, the ownership, benefit and use of data are not necessarily 'bundled' together. The separation of different rights unleashes the potential of distributed innovation while facilitating more subjects to. The separation of different rights unlocks the potential of distributed innovation and facilitates the sharing of the dividends of digital development.

Based on a series of technological innovations such as Ether and NFT, Web 3.0 is finally ushering in a new opportunity for development at present. Conceptually, the concept of Web 2.0 was introduced at the beginning of the twenty-first century under the impetus of O'Reilly Media and others, whose positive affirmation of the importance of the Internet against the pessimistic backdrop of the bursting of the Internet stock market bubble pointed to the important transition of the evolution of the Internet from the 1.0 era as a portal only to the 2.0 era with its two-way interaction and platform-based features. In the logic of the conceptual evolution of 1.0 and 2.0, 3.0 was naturally proposed, but it was not given a substantial connotation for a long time. For example, as early as 2009, Berners-Lee proposed the idea of Web 3.0 and considered the 'Semantic Web' to be his idea of Web 3.0. However, the vague connotation of the Semantic Web did not allow people to clearly understand the essential differences between Web 3.0 and 2.0 and 1.0. This situation only achieved an essential breakthrough in recent years with a series of innovations in distributed technology architectures such as blockchain, Ethereum and NFT. Despite many differences, current discussions about Web 3.0 generally revolve around features such as decentralization and de-trust, in an attempt to achieve a distributed Internet with fewer connectivity constraints and boundary thresholds. As we pointed out in our comparative analysis of the World Wide Web and Solid, the platform-based Internet of the Web 2.0 era is still fraught with the fragmentation and obstruction of 'data shafts'; at the same time, the cyberspace of the Web 2.0 era is more of a 'mapping' of the real space. What Web 3.0 calls for is a rethinking and reform of these two phenomena, which on the one hand requires breaking down the constraints and boundaries of the 'data silo,' and on the other hand is committed to constructing. On the one hand, it calls for breaking down the constraints and boundaries of the 'data silo,' and on the other hand, it aims to construct a new business and a new logic that is different from the real space—just as Bitcoin is an alternative to traditional fiat money. In this sense, we can finally understand that from Web 1.0, and Web 2.0 to Web 3.0, they all follow the same evolutionary logic of pursuing distributed innovation, which undoubtedly provides us with a more macroscopic vision of reform to break through the current plight of the platform-based Internet.

The Never-Ending 'Fork'

The evolution of distributed technologies, from blockchain to Web 3.0, represents the possibility of an 'alternative' Technology Route to the platform-based Internet. However, the new technology types, diversity and complexity make it difficult to cover them all in a point-by-point analysis, so another analytical perspective to reveal the richness of the multiple Technology Routes is to explain the 'methodology' for finding and realizing them. This is exemplified by the 'forking' phenomenon prevalent in the open-source software field.

A 'fork' is a term often encountered in the open-source community and implies a process whereby the community disputes the direction of the code and splits into different versions that are somewhat incompatible with each other. A successful fork will divert some of the community's programmers or even lead to the demise of the current code. But this does not mean that 'forks' are a 'cancer' on the open-source community; on the contrary, it is the prevalence of 'forks' that is the real reason why they are so common. On the contrary, it is the prevalence of 'forks' that sustains the open-source community's drive for growth and innovation. The explanation for this 'uncommon sense' phenomenon needs to be found in open-source software itself.

Open-source software is computer software whose source code can be accessed at will. The differences between open-source software and closed-source software are mainly in the different software production processes and management models. Historically, the story of open-source software began with a printer whose source code Xerox refused to make public in 1980, leaving Richard Stallman unable to modify the print program to meet his needs. This led Stallman to argue that 'all software should be free, not only to be used and distributed freely but also to be modified freely to suit people's needs.'[9] Stallman founded the Free Software Foundation, a non-profit organization, and launched the Free Software Movement (FSM). In contrast, Eric Raymond revised the FSM regarding the laws of software development. He launched the Open-Source Initiative (OSI). In contrast, Eric Raymond modified the FSM from the laws of software development and launched the Open-Source Initiative (OSI) to emphasize the role of a decentralized, participatory open-source software model in driving software innovation and productivity. In Raymond's view, the key to open-source software is not the rejection of 'private property rights' in the form of closed-source software, but rather a centralized, hierarchical model for the production and management of software. As he puts it, '[open-source software] opens the door to competing ideas, creating miracles of software development with unprecedented speed and incredible performance.'[10] It is also in the sense of OSI that 'openness' is associated with 'participation,' 'innovation' and ever-increasing 'efficiency'.

However, why does 'openness' promote 'innovation'? What motivates a programmer to make his or her hard-won knowledge (code) freely available to the community without any financial reward? Equally important is the question of how open-source software can achieve a division of labor and collaboration if there is no hierarchical leader and everyone participates or withdraws as they wish. Indeed, these are the key questions that have 'bedeviled' traditional economic theory about the open-source software phenomenon, and the 'bifurcation' mechanism is the main response to this theoretical gap.

Regarding incentive motivation, 'bifurcation' implies the recognition, inclusion and encouragement of multiple motivations of participants. The 'economic man' assumption of traditional economics places too much emphasis on short-term, direct material benefits as an incentive, thus ignoring both the possibility of long-term or indirect material benefits and the diversity of producers with multiple motivations such as psychological, cultural, belonging and learning. Allowing 'forking' means that open-source software does not presuppose a motivation for participation, but rather, through the direct presentation of source code and the

[9] See Richard Stallman's Personal Site, https://stallman.org/.

[10] Raymond, Eric, The Cathedral and the Bazaar, *Knowledge, Technology & Policy*, 1999, 12:23–49. http://dx.doi.org/10.1007/s12130-999-1026-0.

low or no threshold requirements for access, tacitly acknowledges that programmers may be motivated to participate in the software development and production process based on heterogeneity. During this process, programmers may also leave or 'start a new business' at any time due to a change in motivation. While 'forking' the original code reduces its support and accumulation, the open-source community as a whole, benefits from including diverse motivations, which constitutes a constant source of innovation in open-source software.

Regarding division of labor and collaboration, 'bifurcation' implies an amalgamation of mechanisms such as self-selection, self-organization and the politics of the able. In contrast to the hierarchical system, which emphasizes identity and boundary control, 'bifurcation' is more about the affirmation and maintenance of spontaneous organizational order. In this case, the division of labor to form an organizational outcome no longer implies a place or collection of elements where managers organize different members together to accomplish collective action but is seen as a spontaneous process where different participants deliberate and cooperate to meet their own needs and form an orderly collective action. The 'bifurcation' breaks down all the structural constraints that participants have to face in other organizational contexts. The three options Albert Hirschman wrote about (i.e., withdrawal, appeal and loyalty) that are truly placed in the hands of the producers themselves, in order to ultimately achieve a collective action through 'free dialogue, consensus-based decision-making and small-group alliances based on individual ties,' are the composite mechanisms for joint production.

At both levels, 'bifurcation' means exploring the behavioral and organizational foundations of a diverse Technology Route. Let's take this as a starting point for understanding the platform Internet. We can easily see that the control of the 'bottleneck' resources in the technology system by the platform is in fact the structural constraint that limits the 'fork,' and the progress of Solid compared to the World Wide Web is also in breaking the 'data shaft' to prepare the conditions for the 'fork' at the technical structure level. The advancement of Solid over the World Wide Web is the breaking of the 'data shaft' to prepare the conditions for 'forking' at the technical, structural level. In this sense, 'forking' does not only mean 'competition,' but also the continuous reflection on and rejection of any technological architecture, which is the deeper meaning of exploring multiple Technology Routes.

CHAPTER 5

Global Governance of Transnational Digital Platforms

Digital platforms have undoubtedly become important players and influencers in the current global political and economic system. Many digital platform companies have already surpassed the gross national product (GDP) of most countries in terms of market capitalization size. In December 2021, for example, Apple's market capitalization was already close to $3 trillion, surpassing even the total GDP of the UK in 2020. It is worth noting that this change has only occurred in the last fifteen years, with Apple's market capitalization before 2006 being less than $10 billion. Apple is not an isolated case—Google, Facebook, Amazon and super digital platform companies such as China's Alibaba and Tencent all have market capitalizations comparable to the GDP of the world's major countries. However, the global reach of digital platform companies does not stop there, as they are themselves deeply involved in all aspects of governance processes, both within countries and globally. In particular, in the context of the Newcastle pneumonia epidemic, digital platform companies have played a vital role in global public health governance by their technological and platform ecosystem strengths. Google and Apple have jointly developed mobile apps that can alert 'close contacts,' while China's 'health code' technology system was initially developed by Alibaba Group and rolled out nationwide. But on the other hand,

Y. Jiang and K. Jia, *Transformation of Platform Governance in China*, IPP Studies in the Frontiers of China's Public Policy,
https://doi.org/10.1007/978-981-99-6456-7_5

after nearly two centuries of rapid development in an inclusive regulatory environment, global digital platforms have been facing the pressure and impact of the 'techlash' movement globally since 2016. Countries have begun to reflect on the current rationale and applicability of inclusive regulatory frameworks for global digital platforms, which are also being 'grilled' by national regulators on issues such as digital taxation, fake news, antitrust and unfair competition. Many of the global digital governance issues covered in the first six chapters of this book are, to a large extent, closely related to global digital platforms: on the one hand, the global expansion of digital platforms, while driving the process of digital globalization, has led to the rise of governance issues that were originally confined to local areas into the focus of global governance (e.g., digital taxation); the unique business and organizational models of digital platforms have made them themselves. On the other hand, the unique business and organizational models of digital platforms have made them a major cause of many global governance challenges (e.g., cross-border data flows). It is for this reason that the global governance of transnational digital platforms has become one of the focal points of global digital governance. This chapter, however, will focus more on the global governance challenges arising from changes in organizational models such as 'platforms,' unlike those above domain-specific global governance issues. We need to recognize that one of the major changes brought about by digitization is the emergence of 'platforms' as an important, if not dominant, organizational model of the current era and that the expansive nature of digital platforms makes platform governance naturally global. These considerations form an important part of the discussion in this chapter: the transformation of platforms as forms of organization and the governance challenges and responses to the transformation of platforms.

The Spread of Digital Platforms and the Global Rise of the Platform Economy

It is difficult to say who was the 'first' digital platform, but a process of digital transformation has accompanied the spread of platforms as a form of organization. During the second Industrial Revolution, the machine-made factory was seen as an important form of organization, replacing the decentralized smallholder economy as an important industrial form. It is worth noting that the sociology of technology and industrial organization studies also suggests that the machine-made factory was not the only form of organization at the time and that the 'flexible specialization' model was still important **at** the time. In other words, the second Industrial Revolution did not necessarily lead to mega-factories characterized by the Ford assembly line but also to highly specialized, skilled and decentralized clusters of factories. However, in any case, the factory model of organization, characterized by a hierarchical structure with clear boundaries and a more fixed labor contractual relationship with workers, continued to be the dominant industrial form as the effects of the Industrial Revolution spread.

The new technological revolution has not only brought about more efficient production methods but has also had a significant impact on traditional organizational models, with 'platforms' being a disruption and even a replacement for this. In the 1970s and 1980s, when the development and industrialization of integrated circuits were accelerating, Intel, recognizing that it would not be able to produce all the integrated circuits needed for computers, turned to the central processing unit (CPU) as its core competency and, by defining the interface to the CPU, succeeded in creating a 'platform' business. By defining the interface to the CPU, it managed to create a 'platform' business ecosystem that leveraged other manufacturers to produce components around Intel's CPUs. It is worth noting, however, that although the Intel platform already had important characteristics of a digital platform (mobilizing producers from outside the organization), it was still more akin to the 'standards' of the industrial era and did not yet fully reflect the change in the organization of the industry, which only gradually manifested itself in companies such as Google and Facebook after the commercialization of the Internet had begun. The latter was only gradually reflected in companies such as Google and Facebook after the commercialization of the Internet.

The invention of the World Wide Web kicked off the commercialization of the Internet at the end of the last century, and the emergence of digital platforms was one of the hallmarks of this phase. On the one hand, community-based 'mass production' phenomena such as Wikipedia and open-source software quickly took shape. They significant attention, and for the first time, the huge potential of the Internet to liberate productivity was recognized—that is, through the open network structure, the Internet through its open network structure, the Internet has been able to facilitate the flow of knowledge, while at the same time unleashing the knowledge production capacity of dispersed individuals and ultimately creating a vast array of knowledge products. On the other hand, the voluntary nature of 'mass production' and the difficulty of maintaining a consistent quality of service or product make its transfer to a 'for-profit' commercialization model unlikely. The 'digital platform' organizational innovation, which is based on openness and at the same time improves the management of the production process, has become a new way for entrepreneurs to 'make money' from the Internet.

Napster, which aimed to provide a peer-to-peer music-sharing service, was not the first digital platform, but it can be considered the first to be widely used. Within just four months of its launch in June 1999, Napster had 150,000 registered users, and by February 2001, this figure had risen to 26 million, with even the Bloomberg Business Review suggesting it may have reached a maximum of 80 million users.[1] Napster's business model is to provide a 'platform' for aggregating users to share music between users; it does not provide a music product or service directly to users (similar to a traditional record label), the 'users' on top of the platform are the providers and consumers of that product or service. Although the subsequent development of Napster has been greatly affected by the copyright lawsuit with the record companies, the 'digital platform' is a new form of organization that is gaining popularity among Internet entrepreneurs. Google, founded a year before Napster, formally established itself as a major organizational model in the digital age when it began selling search term advertising as a profit model in 1999. Google aims to provide users with quality search results from a vast amount of information that does not come from Google itself but from the producers of information on the Internet; Google does not charge fees to the suppliers or consumers of information but receives revenue from advertisers to support the operation of the platform. In this way, the digital platform links several actors, such as suppliers, consumers, aggregators and third parties. It maintains a stable and lasting process of transactional services while remaining open. It is important to note that when Google defined the basic operating model of the digital platform in 1999, the focus of Google was still only on the 'information exchange,' and how to maintain the continuous input of information producers (i.e., the information production process before the information exchange) was not part of the consideration of the digital platform at that time. Internet

[1] Eamonn Forde, Oversharing: how Napster nearly killed the music industry, https://www.theguardian.com/music/2019/may/31/napster-twenty-years-music-revolution.

openness was already assumed to be a continuous incentive for decentralized subjects to produce. Despite the later development of user-produced content sites (e.g., YouTube) and the widespread recognition and use of advertising sharing schemes, digital platforms still dominate the amount of revenue distribution throughout the chain of information production and trading and lead to a fragmentation of the power and distribution structures that form one of the focal points of the current rethinking of the development model of digital platforms.

Although many digital platforms, such as Facebook and Twitter, have copied Google's operating model, the advertising-based digital platform structure is only part of platform economy model. As the commercialization of the Internet progresses, digital platforms are increasingly being introduced into various sectors and eventually evolving into a spectacularly rich industry. Nick Snicker, a lecturer at King's College London, classifies digital platforms into five categories according to the different functions they play, including advertising platforms (e.g., Google as a search engine), cloud platforms (e.g., Amazon as a provider of cloud services), factory platforms (e.g., GE's Industrial Internet Platform as a provider of transformed production management processes), product platforms (e.g., Spotify as a provider of music products to users) and transaction aggregation platforms (e.g., the YouTube-based platform that matches drivers and passengers). The Rise of the Digital Platform Enterprise: A Global Survey, published in 2016 by Dr. Peter C. Evans of the US-based Platform Strategies Institute and Annabelle Gawer, Professor at the University of Surrey, UK, categorizes digital platforms into four types: trading platforms (e.g., Youbou), innovation platforms (e.g., Jitterbug), investment platforms (e.g., Alibaba Holdings) and integration platforms (e.g., Apple iOS App Store). These studies reflect the diversity and complexity of digital platforms themselves, as well as the breadth and depth of their use in various sectors.

However, the impact of digital platforms is not limited to the impact on the organizational models of individual companies; it is more important and far-reaching in terms of its impact on the transformation of the industrial ecosystem. Studies have shown that, similar to the conglomerate model developed in East Asian countries after World War II (e.g., Japanese conglomerates, Korean zaibatsu), digital platform companies have chosen to expand horizontally or vertically, using their data or capital advantages to form platform business conglomerates, thus extending their influence into different vertical industries. For example, as an e-commerce platform, Alibaba has formed a payment platform, a logistics platform and a computing platform and based on these, has further expanded into various areas such as social media, lifestyle services and mobile travel. As a search engine platform, Google has a business scope that also covers different areas such as online video, social media, mapping and autonomous driving. The cross-networking effect of platforms drives the expansion of digital platform companies as a flexible and open form of organization, which allows them to leverage on the strength of the users they have already gained to leverage other market segments. This is why digital platforms have a stronger incentive to expand and are less constrained to do so than traditional forms of corporate organization. It is therefore important to summarize the unique characteristics of digital platforms as an organization and, on this basis, to understand the far-reaching implications of the global expansion of international digital platforms.

The Transformative Significance of Digital Platforms as a Form of Organization

Although the transformative significance of digital platforms can be summarized from different perspectives, they represent the development and transformation of how productive human life is organized. In her 2017 book *Law and the Wealth of Nations: Finance, Prosperity and Democracy,* the late Tamara Lothian, a professor at Columbia University Law School, argues that the 'most advanced' at different stages of history. 'Productivity' at different stages of history does not refer to the most efficient (i.e., highest input–output ratio) mode of production, but rather to a mode of production that goes beyond what is already in place and unleashes the greatest degree of human creativity. In the industrial age, this was manifested in assembly-line production based on the division of labor (e.g., the 'Ford system'); in the digital age, it refers specifically to the economies of scale that result from the mobilization of decentralized agents to produce.[2] According to Professor Rossien's definition, the progressive significance of digital platforms is precisely their mobilization of the 'productive initiative of decentralized agents' and, as a

[2] Tamara Lothian, *Law and the Wealth of Nations: Finance, Prosperity and Democracy*, New York: Columbia University Press, 2017.

result, their 'better' unleashing of human creativity than the Ford system of large factories. Given the wide range of human productive activities, this progressive significance of digital platforms can be summed up in three areas: the construction of bilateral markets, the shaping of public space and the deepening of open innovation.

(1) The 'non-neutrality' of platforms in influencing price structures from a bilateral market perspective. Another concept in economics that is closely related to digital platforms is the 'two-sided market,' which refers to a market in which both buyers and sellers are present and bidding at the same time, as opposed to the 'one-sided market,' is a market in which only buyers or sellers bid due to an unbalanced market structure. The digital platform is the vehicle and space for buyers and sellers to match each other in a bilateral market. On the face of it, bilateral markets seem to cover a wide range of situations. For example, free trade farmers' markets are often regarded as bilateral markets, but it is clear that farmers' markets are not yet organized as 'platforms,' which raises the question of what the nature of platforms is as a form of market organization.

In his seminal paper with his co-workers, 2014 Nobel Prize winner Jean Tirole summarized three elements of the 'platform economy': the existence of two or more sets of customers (e.g., drivers and passengers), the existence of network effects between different customers (e.g., the more drivers on a platform, the easier it is for passengers to get a ride on that platform) and the need for intermediary involvement (i.e., the need for a platform to match drivers with passengers).[3] David Evans, one of the first scholars of platform economics at the University of Chicago Law School, summarizes the meaning of 'platform' similarly.[4] These three elements may theoretically clarify what a 'platform economy' is. However, they fall short of the key to what a 'platform economy' is in terms of an industrial organization model. In other words, what role does the 'platform' play in trading? Farmers' markets can also play a role in brokering transactions, so can they be called 'platforms'? If the answer is yes, then the 'platform economy' does not seem very different from classical economic theory. Tyrrell concludes his article by explaining that the key

[3] Jean-Charles Rochet and Jean Tirole, Platform Competition in Two-sided Markets, *Journal of the European Economic Association*, Vol. 1, No. 4 (June, 2003), pp. 990–1029.

[4] David Evans and Richard Schmalensee, *Matchmakers: The New Economics of Multisided Platforms*, New York: Harvard Business School Press, 2016.

to a model of organization being called a 'platform' is the 'non-neutrality' of the transaction, i.e., the fact that the seller cannot fully pass on the fees charged by the platform to the buyer; in other words, a 'platform' exists because it can effectively employ cross-subsidization strategies and can impose different pricing strategies on and influence sellers and buyers differently. A market that satisfies this condition is called a 'bilateral (or multilateral) market,' as opposed to the 'unilateral market' of traditional classical economics, and this leads to the main difference between 'platform economics' and traditional classical economics, namely the specificity of the 'price structure.'

Under traditional economic theory, the equilibrium price in a perfectly competitive market is equal to marginal cost. In contrast, for an imperfectly competitive oligopolistic market, the equilibrium price satisfies the Lerner equation, which states that the difference between price and cost is inversely proportional to the elasticity of demand. For platform economics, however, one of its main conclusions is that the pricing of profit-maximizing products is not directly related to marginal cost but rather to price structure rather than the price level. Thus, platform economics parted ways with traditional economics. Take the example of a mobile app shop. Developers upload applications to the platform, and users purchase the related programs. The app shop can charge both the developer (the seller) and the user (the buyer), the sum of which constitutes the 'price level' of the platform, and the relative ratio of the two is known as the 'price structure.' The analysis by Tyrrell et al. shows that the price charged to either side (buyer or seller) in a profit-maximizing platform does not depend directly on variable costs (and thus differs from conventional economics) but is proportional to the bilateral elasticity of demand. This leads us to an important characteristic of platform economies (and bilateral markets): indirect network effects (sometimes called 'cross-network externalities').

Indirect network effects are those where the size of the network on the one side (e.g., the buyer) will impact the size of demand on the other side (e.g., the seller). For example, the more apps there are in an app shop (seller size), the greater the demand from users; conversely, the greater the number of users of an app shop (buyer size), the more programmers will be willing to develop apps on that app shop platform. However, it is important to note that the size of the indirect network effect (i.e., the cross-demand elasticity) may differ for the bilateral parties (buyers or sellers). In other words, the 'seller's influence on the buyer' may be differ from the 'buyer's influence on the seller'; because of this, the price charged by the platform to both parties may be differed. Take the example of online taxi platforms. The elasticity of demand of passengers for online taxis is smaller than that of online taxis for passengers, so platforms tend to heavily subsidize passengers to attract buyers to join the platform (which means that the platform imposes 'negative pricing' on passengers) while charging a commission to the drivers of online taxis. On the other hand, as user habits develop, passenger demand for online taxis gradually becomes more elastic, so platforms can gradually remove subsidies (which means they gradually raise prices for passengers) to make more profit.

The transformative significance of platforms in the context of bilateral markets is thus clear: the 'non-neutrality' of their influence on different market players that them to create a cross-network effect and, as a result, gain superior power over market players. Especially in the digital age, the expansion of digital platforms further amplifies this 'non-neutrality,' enabling them to gain a large number of users in a short period and use this as a basis to leverage more ecological players, thus forming a positive cycle. Digital platforms thus gain a 'first-mover advantage' and enjoy a 'win the digital platform' thus gains a 'first-mover advantage' and enjoys a 'winner-takes-all' development dividend.

(2) The 'reliability' of digital platforms from the perspective of public space. The 'non-neutrality' of digital platforms from the perspective of bilateral markets is an explanation of the transformative significance of platforms based on economic theory. However, the expansive and pervasive nature of digital platforms has extended their influence beyond the economic sphere and into the political sphere. The role of digital platforms in the early days of the commercialization of the Internet was mainly to change and replace the medium of information interaction. The structure of information production and circulation established during the Industrial Revolution, based on mass media such as newspapers, radio and television stations, has been transformed by the impact of the Internet into an end-to-end flat and direct interaction, where anyone can become a producer of information and the transmission of information no longer needs to be regulated by the mass media as a 'gatekeeper.' However, this does not mean that 'control' has disappeared. As Harvard Law School Professor Lawrence Lessinger pointed out at the turn of the last century, the Internet is not the 'freedom' that the 'utopians' had envisaged. The rise of digital platforms has also shaped the new 'rules' of cyberspace, influencing the interaction of information in human society while creating a new 'architecture' for the production and flow of information.

It is important to note that it is difficult to make a direct positive or negative judgment on the impact of digital platforms on the structure of information interaction in human society, which depends on the choice of business strategies and technological solutions of the digital platforms themselves, as well as on the development and evolution of the information environment in which human society as a whole is living. However, putting aside the debate on specific issues at the micro-level, the question of how to explain the political changes brought about by the new structure of digital platforms as a whole remains a major issue that needs to be addressed. In this regard, the analysis from the perspective of the evolution of public space is instructive.

Public space has always been fundamental to the discussion of basic political rights in human society. However, public space itself has evolved to a large extent due to technological and business developments. Dr. Zuo Yilu of Peking University Law School, in his book Beyond the 'Street Corner Speaker': The Margins and the Centre of the Right to Expression, reviews the history of the evolution of information dissemination models in human society.[5] On the one hand, the traditional and classic model of the right to political expression is the 'street corner speaker' model proposed by Owen Fiss, a leading American constitutional scholar and professor at Yale Law School, which depicts a typical scene of people standing on soap boxes on urban street corners and expressing their political views to a crowded audience. The 'urban street corner' constitutes a public space for people to communicate their political views and an important support for the institutional arrangements that influence human society. In the 'street corner speaker' model, the information dissemination occurs directly between the 'speaker' and the 'listener,' with a flat structure that avoids the involvement and influence of any intermediate media. On the other hand, with the rapid development of media technologies such as newspapers, radio and television, from the 1940s and 1950s onwards, humanity entered the era of 'mass communication.' American political scientist and communicator Harold Laswell systematically introduced the concept of 'mass communication' in his article 'The Structure and Function of Communication in Society,' adding the term

[5] Zuo Yilu, Beyond the 'Street Corner Speaker': The Margins and the Centre of the Right to Expression, Peking: Social Science Literature Press, 2020.

'medium' to the 'street corner speaker' model. The role of the media was added to the model of the 'street corner speaker,' which led to a significant reorientation of public space. The media became an important factor in influencing society's political, economic and social structure of society. Along with this process, public space was transformed from the figurative 'urban street corner' depicted by Firth to a 'new paradigm' constructed by the media, as Laswell had observed. As Dr. Zoe Yilu points out in her book, Firth himself had already proposed in his 1995 article the end of the 'street corner speaker' model and the need to focus on the 'new technological scenario.'

Firth's sensitivity to the dynamics of technological development can also be used to guide the analysis of change around digital platforms. However, the latter is far more complex than traditional media such as newspapers, radio and television. On the one hand, digital platforms still exist as media in their own right, constructing the public space of online opinion in the digital age, reaching a much larger group of users by their ubiquity and openness, and because they have not involved in the production of information themselves, appearing to facilitate the interactive processes of human society in a freer and less intrusive manner. However, on the other hand, as Reisinger points out, the openness of digital platforms does not mean 'unregulated (or non-interfering),' but rather 'regulated' (or interfere). According to Professor Hu Ling of Peking University Law School, 'openness' means 'architecture' that can be changed, customized and produced. In other words, although digital platforms have stripped away the information production functions of traditional media and even provide information filtering functions in a more neutral and user-preferred manner to a large extent, this does not mean that information filtering, selection and other communication management functions do not exist in the public space constructed by digital platforms. Both the digital platforms themselves and the ecological actors on top of them may perform communication management functions and shape the public space's public opinion field. For example, the social media bots mentioned in Chapter 5 of this book, entitled 'Algorithmic Global Governance,' can also be seen as a manifestation of communication management functions and may eventually influence major political events such as the UK's EU referendum or the US elections.

If the change in the mass media from the 'street corner speaker' model lies in focus on the role of the 'medium,' then how to understand the change in the public space of human society (especially the public space of political opinion) in the context of the rise of digital platforms remains a matter of debate. The question of how to understand the transformation of the public space of human society (especially the public space of political opinion) in the context of the rise of digital platforms remains a matter of debate. However, as the scale of digital platforms has increased, the debate has begun to show a certain consensus trend. More and more stakeholders are now recognizing the 'reliability' of the public spaces constructed by digital platforms. This public space is now more inclusive and extensive than the 'street corner speaker' model and can even instantly create transnational arenas of opinion; on the other hand, it is also more open and innovative than 'mass communication.' On the other hand, it is also more open and innovative than mass communication, as more and more ordinary users can participate in information production and dissemination, which eventually evolves into a competitive situation of 'crowded voices.' The opposite of inclusivity, breadth, openness and innovation is the more secretive and efficient 'regulatability.' Regulated public space is ultimately the key feature of the transformation of social organization driven by digital platforms.

(3) Digital platforms and 'open innovation' from a production process perspective. Both the bilateral marketplace and the change in organizational form from the perspective of public space were important business forms that received widespread attention in the early stages of the development of digital platforms such as the rise of e-commerce and search engine platforms, and the spread of social media platforms. In contrast to these two types of platforms, the emergence of digital platforms with a production process management function did not attract more attention and focus until the emergence of the 'digital odd-job economy,' which began with the founding of YouTubers in 2009. Although open-source software and Wikipedia can be seen as typical examples of the Internet's liberation of productivity, it was difficult to transform this model into an effective business model for a long time, one of the main reasons being that an open production model based on voluntary participation was difficult to maintain a stable and sustainable level of production and to meet the requirements of a system of product (or service) responsibility regulation. It was not until the rise of the 'digital odd-labor economy' that digital platform emerged as managers of decentralized production processes eventually transposing the Wikipedia model of 'open innovation' to the business world.

Although open innovation is credited in academic thought to Henry Chesbrough at the University of California, Berkeley, its core has long been nurtured in the practice of open-source software and Wikipedia, both of which are often referred to by Harvard Law School Professor Youchai Benkler as the 'Peer Production' model. Harvard Law School Professor Youchai Benkler often refers to the latter two as the 'Peer Production' model. Unlike the Ford system of industrial mass production, 'peer production' emphasizes the importance of decentralized individual participation in the production process and achieves economies of scale through diverse incentives and open organizational models. Take open-source software as an example. The freely available source code allows anyone to 'reproduce' on this basis, thus creating a vast knowledge base of code and laying the technological foundation for the digital economy. From the point of view of productivity theory, the importance of 'mass production' is that it has become the representative of the 'most advanced productivity' in the digital age. Along with the commercialization of the Internet, the 'mass production' model has gradually spread from open-source software and Wikipedia and has contributed to the emergence of the 'digital odd-job economy.'

Marked by the establishment of Uber in 2009, more and more digital platforms began to transform the service sector, with the taxi industry and the hotel industry leading the way and gradually developing into different verticals such as takeaways and lifestyle services. For these industries, the traditional labor model based on geography or acquaintanceship limits the labor participation of non-local, non-acquaintanceship dispersed labor. However, with the ability of digital platforms to aggregate transactions, this threshold is greatly reduced, allowing anyone across geography and language to participate in the production and benefit from it; online taxi and takeaway services are similar, and their value is also reflected in lowering the threshold of production participation to. This basic logic is in line with open innovation such as open-source software and Wikipedia. But on the other hand, the 'digital odd-labor economy' is always faced with the organizational challenge of coordinating the production processes of dispersed individuals to achieve economies of scale. In the case of 'mass production,' which relies more on decentralized, self-organized online communities, the organization process is fraught with conflict, change and instability, even though the output is also highly productive (e.g., large and rich open-source software code and Wikipedia entries). In contrast, the current process of commercialization of the 'digital odd-job economy' has forced it to rely on platforms such as transaction facilitators, process managers and output controllers, and as a result, has partially lost its traditional characteristics of decentralization and self-organization, and has become more subject to platforms.

Based on this analysis, it is possible to understand the digital platforms' transformative significance from the production process management perspective. Open innovation is a common feature of both the 'mass production' model and the 'digital odd-job economy,' but what is different is the process and how they organize decentralized production. The former embodies the progressive significance of lowering the barriers to production and unleashing the creativity of dispersed workers, accompanied by the flexibility and instability of self-organization; the latter inherits the labor model of dispersed workers but is more organized and controlled to maintain consumer-oriented commercialization, and the institution that enables this organizational control is the 'digital platform.' The latter, while inheriting a decentralized labor model, is more organized and controlled to maintain a consumer orientation, and the institutions that fulfill this organizational control function are the 'digital platforms.'

It is clear that while recognizing the positive aspects of digital platforms inheriting the characteristics of 'open innovation,' their control and management of the production process will inevitably lead to criticism and attacks from the governed. Reflections on takeaway platforms where delivery workers are 'trapped' in the system, and criticism of Apple's app shop for taking excessive commissions that squeeze the income of innovators, have been the focus of global attention in recent years. In understanding the implications of the changes brought about by digital platforms, a further analysis naturally involves the governance risks and challenges accompanying this change process. These challenges and risks are naturally global due to the global expansion of the commercialization of digital platforms.

Governance Risks and Challenges of Digital Platform Change

The transformation of digital platforms as forms of organization has brought not only new developments, but also new governance risks and challenges. The three characteristics of digital platforms—'non-neutrality,' 'regulatability' and 'open innovation'—encompass both positive and negative effects. While recognizing that digital platforms can effectively mobilize and organize decentralized productive forces, thereby achieving economies of scale and scope, we cannot ignore the negative effects that the change process may have on individual rights or the public interest. It is worth noting that although digital platforms are mainly involved in economic activities and are not yet commonly used in human social and political activities, from the perspective of global governance, the extensibility, connectivity and openness of digital platforms themselves make them a source of governance risks across economic, social and political spheres. In its Report of the Working Group on Platform Scale: White Paper, published in October 2020, the Stanford University Center for Network Policy pointed out bluntly that the governance risks and challenges posed by digital platforms are not limited to the antitrust arena, but also it has far-reaching social and political implications. For this reason, this section will summarize the global governance risks and challenges of digital platform change from three perspectives: economic, social and political.

It is important to note that, unlike the problem-specific analyses that conducted, this section focuses on the general causes of governance risks on digital platforms. In other words, the basic argument in this section is that there are general causes and inherent patterns in the decentralized governance issues arising from digital platforms. The response at the governance level should be based on a holistic reflection and consideration, in addition to a targeted approach of 'treating the head when the headaches and treating the foot when the foot hurts,' so that a systemic response is possible beyond the problem-specific 'governance dilemma.'

(1) The data 'shafts' of the platform economy and the economic risks of monopolistic market structures. The governance risks of digital platforms in the economic sphere basically revolve around their monopolistic market position. In existing research and practice, first-mover advantage, network effects and 'winner takes all' are seen as natural attributes of digital platforms, and this has largely influenced regulators' perceptions of the laws governing the development of platform economies. Digital platforms are seen as market infrastructures, and the resulting monopolistic market structure of the platform economy is seen as economically justifiable. However, as digital platforms grow in size and influence, more and more countries are becoming concerned about the 'abuse' of monopolistic market positions by digital platforms and, in general, the extent to which the monopolistic market structure of the platform economy will hinder the prospects for sustainable innovation in the digital economy in the future. In the face of these questions, research and practice have gradually identified the 'territorial' movement of digital platforms toward user data as the root cause of governance risks and proposed a series of solutions including technological innovation and institutional reform, which constitute a systematic response to the anti-monopoly governance of the platform economy.

For a long time, national regulators have focused their research on the antitrust aspects of digital platforms on the specificities of their development models and the challenges they pose to the established antitrust legal system. For example, in December 2014, the then-newly appointed President of the EU Competition Commission, Margrethe, announced that she was postponing a final decision on the Google antitrust investigation. She said that the Google antitrust case was 'a complex issue involving many parties and, while the rapidly evolving market requires us to make decisions quickly, I have to take more time and evidence to consider the next steps.' The EU's antitrust investigation began in 2010 with a joint lawsuit by 18 European companies against Google, and in March 2013, the EU antitrust authorities published their preliminary findings, declaring that Google had allegedly abused its monopoly power in four areas. First, the display of search results. Google allegedly deliberately removed search results from competitors (i.e., other search engines) and prominently placed paid advertisements on search pages without the knowledge of users. Second, the issue of content grabbing. Google allegedly cited original content from third-party websites in its products without the consent of the originators, thereby causing a significant drop in traffic to the original websites and hindering the development of SMEs. Third, the issue of mandatory agreements with third-party media. Google is suspected of forcing third-party media (e.g., newspapers, publishers, portals) to display only Google ads or Google search results on their homepages, thereby squeezing the market space of other search engines. Fourth, the issue of compulsory advertising contracts with commercial companies. Google allegedly compels commercial companies to publish online advertisements only through the Google platform and not to publish or promote them on other search engines simultaneously.

However, these four allegations do not conclusively establish that Google has violated the relevant antitrust laws. This is partly due to the lack of data to support the conclusions of the antitrust investigation body but also because the Internet platform companies represented by Google have a special nature that is very different from the traditional industrial giants, which makes it difficult to adapt the antitrust theory born in the industrial era and the legal system built on top of it to the rapidly developing Internet industry, which constitutes the current antitrust dilemma of digital platforms. The 'dilemma.'

The traditional antitrust doctrine was born from the industrial era and was designed to restrict the association of firms that affected competition in the marketplace. For this reason, the US Antitrust Act is also known as the 'Antitrust Act.' The control of production by firms to maintain monopoly profits, resulting in an inadequate supply and high prices and a corresponding loss of efficiency in the allocation of resources, is the reason for antitrust legislation. Over time, antitrust law has developed a set of economic tools for determining whether an enterprise is suspected of abusing a dominant position in a market, namely the 'three-stage theory': defining the relevant market for the product in question, determining whether the enterprise being sued constitutes a dominant position within that relevant market, and if so, whether the enterprise has abused its market control. Of these three steps, the definition of the relevant market is the starting point and basic premise of the competition analysis, which is traditionally determined through methods such as the hypothetical monopolist test (SSNIP). The SSNIP determines the scope of the relevant market by assuming that the price of a product has increased and that consumers have chosen alternative goods. However, all types of digital platform services, including search engines, are essentially free to consumers. In other words, there is no price-based transactional behavior between supply and demand on search engine platforms. There is no question of changes in the number of transactions due to changes in transaction prices, and market definition is therefore not possible. However, even assuming that there is a price for Internet services, it is still difficult to define the relevant market precisely because digital platform companies, including Google, are typically 'bilateral markets.' A bilateral market means that there are three parties involved in the transaction: the platform that provides the service (e.g., Google), the user who receives the service (e.g., the consumer) and the third party that accesses the platform (e.g., the web content provider). The utility of the user depends to a large extent on the prosperity of the third party—in other words, whether consumers choose to consume alternative goods depends not only on the elasticity of supply and demand in a unilateral market but also on the cross-elasticity between bilateral markets, the latter of which is not currently measured by suitable means.

If the current maladjustment of the antitrust legal system was limited to the technical aspects mentioned above, the problem would not be too serious. With improved research methods or better data, we could well solve the problems of defining the relevant markets and calculating the cross-elasticity of bilateral markets. However, the impact of digital platforms goes far beyond this. The uncertainty created by rapidly changing technological innovations and increasingly complex business models is the crux of the problem. Traditional antitrust law is essentially a 'hierarchical' system of governance based on a 'principal-agent' relationship. The principals (the legislator versus the regulator and the higher regulator versus its subordinates) have a clear regulatory objective and fully understand how to achieve that regulatory objective. With appropriate resources and accountability according to established rules, the delegator can motivate other participants—i.e., agents—to achieve the goal. However, the problem now is that in the face of rapidly changing technological innovations and increasingly complex business models, principals cannot give clear goals and rules, and agents need help achieving regulatory enforcement within the established framework. The uncertainty of the Internet business landscape constructed by digital platform companies is particularly acute because even Google itself, which best understands its technological properties and business model, has no idea of the possibilities for the future development of the Internet. Otherwise, it would not need to watch only as Oracle, having misjudged the prospects for the adoption of Java; otherwise, it would not have had to watch Oracle take a slice of the huge profits in the Android market, or it would not have had to do anything about the emergence of Facebook and Twitter because it ignored the social media market. Indeed, digital platforms have often cited this 'uncertain' future competitive challenge as an important reason for not occupying a monopolistic market position in response to antitrust pressure from regulatory authorities, arguing that excessive antitrust pressure would prevent both innovation and fair good market competition.

However, regardless of the controversy, antitrust investigations against digital platforms worldwide have intensified in tandem with the expanding scale and scope of digital platform companies. The EU's antitrust investigation against Google in 2010 was only a 'wave' in a global wave of antitrust regulation of digital platforms. According to the 'Platform

Economy Antitrust Regulation Observation Report (2021)' released by the Nandu Antitrust Research Group in December 2021, the four major digital platforms—Google, Apple, Facebook and Amazon—were involved in a total of 149 monopoly-related cases from 2017 to 2021, including 34 new cases in 2021. Google is involved in one-third of the cases and has been fined nearly US$10 billion, of which more than US$9 billion is from the EU; Facebook is involved in only 21 cases but has also been fined a total of US$5.14 billion. In terms of initiators of antitrust investigations, the UK and EU have the highest number. At the same time, India, Turkey, South Korea, Brazil and Russia have all launched antitrust investigations against large digital platform companies in 2021. Even in the US, where digital platform companies are headquartered, domestic antitrust investigations against digital platform companies continue to be frequent. It is worth noting that 2021 is also known as the 'inaugural year' for anti-monopoly in China's digital platforms and that the Central Economic Work Conference held in late 2020 listed 'strengthening anti-monopoly and preventing disorderly capital expansion' as one of the eight key tasks for 2021. Since then, 92 cases of anti-monopoly enforcement on digital platforms have been enforced, relevant legislative rules have been improved, and an anti-monopoly regulatory system with Chinese characteristics has gradually been formed. In terms of specific cases, the Alibaba anti-monopoly and Meituan anti-monopoly cases are typical of them. The anti-monopoly regulators found that Alibaba and Meituan abused their dominant market position and imposed fines of RMB 18.2 billion and RMB 3.4 billion, respectively, in response to the business strategy of digital platforms requiring participants in economic activities dependent on the platforms to 'choose one over the other.' According to research conducted by the Nandu Antitrust Research Group, after the anti-monopoly penalties against Alibaba and Meituan, the number of digital platforms requiring merchants to 'choose one over the other' has decreased significantly.

However, it is worth noting that 'two-for-one' is only one of the tactics digital platforms use to 'abuse' their monopolistic position in the market. 'Chokehold mergers,' algorithmic collusion, search discrimination, user channeling and unreasonable tariffs are more complex and difficult to define platform strategies. While specific analysis of specific problems can improve the rule system of anti-monopoly regulation of the platform economy, it may be difficult to achieve the ultimate goal of maintaining the sustainable development of the digital economy in the future due to the lagging nature of legal rules and the uncertainty of the evolution of platforms. In this context, going back to the 'root causes' that led to the formation of monopolistic market positions in digital platforms and exploring possible systemic solutions on this basis may be a new way of thinking different from the existing governance reforms. In the case of the latter, the 'data shaft' has been identified as the key to this and has attracted the attention of researchers and practitioners alike.

Timothy John Berners-Lee, the inventor of the World Wide Web, pointed out that the key innovation of the Web was twofold: the grafting of hypertext technology onto the Internet to allow more types of information, such as images, videos and audio, to be distributed on the Internet; and the use of one-way links, which allowed anyone to link to a resource without any response from the owner. This maximizes openness (i.e., open links are the default) by allowing anyone to link to the resource without any response from the owner. This technical solution, which focuses primarily on 'openness of links,' has had the positive impact of 'openness for innovation,' but has also become the technical basis for the possession and control of data by digital platforms. Specifically, the web-centric technical architecture of the World Wide Web does not separate applications from data, which makes it difficult to share user data openly, as it is split between different web pages (and the companies behind them). Berners-Lee likened this situation to a 'data shaft' (Silos). The 'data shaft' allows the first-mover digital platform companies to enjoy network effects and monopoly power by their user data and also allows them to easily generate huge revenues through advertising models, which is the key reason for the first-mover advantage, network effects and 'winner takes all' of digital platforms. In particular, the 'data shaft' phenomenon also has a 'horse-trading effect.' Once the first-mover digital platform companies enjoy the 'data shaft' dividend, they will further expand their user base and solidify their 'data shafts,' thus eventually forming a monopolistic market structure. From this perspective, the key to solving the anti-monopoly problem of digital platforms is naturally to break the 'data silo' phenomenon in order to unleash the digital economy's shackled innovation. In terms of concrete implementation, the 'right to data portability' established by the EU, the new Solid Internet Protocol actively promoted by Berners-Lee and the much sought-after Web 3.0 concept in 2021 can all be regarded as important paths to break the 'data silo.' The 'data silo' can be seen as an important path to break down.

All in all, under the questioning of the 'data shaft' concept, the monopolistic market structure promoted by digital platforms should be seen more as a specific structure catalyzed by a specific technological solution (e.g., World Wide Web protocols), a specific business model (e.g., advertising models) and not necessarily as an inevitable outcome or a general rule of the evolution of the platform economy. The impact of the concept of 'data shafts' is not limited to the economic sphere but can also be used to analyze the far-reaching impact of digital platforms in the social sphere. For example, issues such as privacy violations and social polarization, which have received widespread attention, can also be explained and responded to more systematically under the framework of the 'data shaft' concept.

(2) The social risks of the evolution of platform capitalism: 'surveillance capitalism' and 'Taylorism 2.0.' The impact of digital platforms is not limited to the economic sphere, but their profound influence on civil rights, values and social structures is also a major concern. Although digital platforms are often created with economic interests in mind, their unique development model makes it inevitable that their impact extends beyond the economic sphere to the social sphere. Slavoj Zizek, a Slovenian philosopher, offers a thoughtful reflection and illustration. In Zizek's view, the great achievements of digital platform companies should be seen in part as based on 'privatizing what Marx called our commons.'[6] Specifically, Bill Gates' great wealth is not due to the higher quality or lower cost of Microsoft's products but rather to the creation of a new medium that enables people around the world to communicate, and to the persistent high 'rents' that he receives from controlling that medium. It is worth noting that the value of the medium does not derive from the value of individual use but from the communal value of shared use by society as a whole. In this sense, Zizek argues that many digital platforms, including Facebook, Google and Amazon, are in essence, 'rent-seeking' in privatizing the social commons in order to realize commercial benefits. For different digital platforms, the 'social commons' they rely on can be broadly understood as the accumulation of human public knowledge or relationships and manifests itself in different forms in different contexts: social media platforms in the form of social connections between users and search engine platforms in the

[6] *See* Slavoj Zizek, We Need a Socialist Reset, Not a Corporate 'Great Reset' https://jacobinmag.com/2020/12/slavoj-zizek-socialism-great-reset.

form of users' preferences for searching information; even in the case of e-commerce and travel platforms, which are directly based on market transactions, the platforms can still improve the efficiency of matching based on the 'social commons' of transaction history data and thus earn 'rents.'

Zizek's 'privatization of the social commons' is exactly what Tim Berners-Lee calls the 'data shaft'—a 'data shaft' created by the 'privatization' of the commons, where digital platforms realize their economic benefits. The 'privatization' of data as a common creates a 'data shaft' in which digital platforms realize their economic benefits. In the process, platform capitalism, primarily profit-seeking, evolves into other forms and has an erosion and impact on social interests.

On the one hand, the business model of digital platforms based on 'data shafts' will continue to demand the generation and accumulation of data, which will inevitably violate users' data privacy rights and even evolve into what Harvard Professor Shoshana Zuboff calls 'Surveillance Capitalism.'[7] The inherent value of data and the uncertainty of its release have led to a tendency for digital platform companies to increasingly collect, manage and control data, and to derive direct benefits from it (in the form of advertising, for example). Data has naturally become a 'moat' for digital platform companies to maintain their monopoly power and a 'shield' to control key aspects of the platform. The data has naturally become a 'moat' for digital platform companies to maintain their monopoly power and a 'talisman' to control key aspects of the platform ecosystem. On the other hand, and perhaps more importantly, the business model of valuing and financing digital platforms based on the data they control provides a significant incentive for digital platforms to accumulate and control data for financial gain, even if they cannot directly profit from the data they control. Indeed, the prevalent combination of venture capital and digital platforms in the digital business environment has allowed the uncertainty of data value to be successfully translated into data-based business valuation and thus the realization of value. This business model has not only been shaped into the 'Silicon Valley model' and is highly regarded, but it has also spread to countries worldwide to become a universal digital business template, which will further incentivize digital platforms to try to capture, own and control data. The logic of valuing

[7] Shoshana Zuboff, *The Age of Surveillance Capitalism, The Fight for a Human Future at the New Frontier of Power*, New York: Public Affairs, 2019.

companies based on the value potential of the data rather than the actual profit that the product may generate has made digital platform companies far more valuable than other companies in the market and the top players in the capital markets.

On the other hand, digital platforms, while promoting 'open innovation,' may evolve into 'Taylorism 2.0' in order to achieve finer and more flexible control over workers and ultimately 'trap' them in the system. The system is 'trapped' in the workforce. Unlike traditional forms of corporate organization, the openness of digital platforms can still effectively mobilize decentralized workers. However, they strengthen the organization and control of the production process through mechanisms such as algorithmic scheduling, the recording of labor behavior, the accumulation of historical credit and the evaluation of results. Similar to the Taylorism of the Industrial Revolution, digital platforms can also achieve deeper and more comprehensive control over workers through technological innovation in algorithms and platform management (e.g., recognizing workers' facial expressions to determine whether they are serious about their work), and eventually evolve into an upgraded version of Taylorism 2.0. This is an updated version of Taylorism 2.0. Just as the behavioral sciences have long reflected on Taylorism's erasure of workers' subjectivity, digital platforms still risk alienating workers in the process of implementing Taylorism 2.0. More importantly, this uneven power structure between digital platforms and decentralized workers will also largely determine the polarization of income distribution structures. According to Professor Martin Kenney of the University of California, Davis, there are, in fact, four categories of people in the ecosystem constructed by digital platforms: the owners of the platforms and the venture capitalists associated with them (e.g., the founders of social media platforms), the head producers of the platform economy (e.g., the netizens of video sharing platforms), the ordinary workers of the platform economy (e.g., the workers of travel platforms), the workers of the platform economy. (e.g., the average driver on a travel platform) and the average user of the platform economy (e.g., the user of a search engine or social media platform). The number of people in each of the four categories increases, but their returns to the platform economy decrease, creating a highly concentrated 'pyramid' structure. According to a 2012 study by Nobel Prize winning economist Joseph Stiglitz, the ratio of CEO to worker earnings in the US marketplace has expanded from 46:1 to 331:1 in the last 20 years.[8] Although Stiglitz's study does not break down the exact ratio of CEO to worker earnings in digital platform companies further, given the rapid growth of digital economy, and the reality of their increasing share of the economy as a whole, we can still infer that the income ratio of digital platform owners to the average worker in the platform economy is likewise likely to be in a constant trend of widening. Professor Mariana Mazzucato of the University College London and Professor Guy Stein of the School of Oriental and African Studies at the University of London have directly criticized the evolution of the platform economy into a 'rent-seeking economy' in the digital age.[9] The evolution of digital platforms from organizers of 'open innovation' to practitioners of 'Taylorism 2.0' thus entails social risks not only in terms of the alienation of individual workers' labor processes but also in terms of the fragmentation and intensification of social inequalities, the latter being the subject of the current era. The latter is the theme of the current era. In this sense, the social role played by digital platforms will become even more prominent and important.

(3) The political risks of 'freedom of expression' and 'technological answers' on digital platforms. 2018 saw the outbreak of the Cambridge Analytica affair, which largely marked the beginning of the political risks of digital platforms. The political risks of digital platforms have become an important issue of concern for all countries. Although Facebook, as a platform, can claim that it does not directly collect user data for political propaganda, its open platform structure still provides a space and venue for third-party actors, including Cambridge Analytica, to manipulate political opinion, and Facebook itself was fined $5 billion for the Cambridge Analytica incident.

It is worth noting, however, that the current strong regulatory attitude toward the influence of digital platforms on political opinion is not 'always the case.' When digital platforms were first established in the late

[8] J. E. Stiglitz, Macroeconomic fluctuations, inequality, and human development, *Journal of Human Development and Capabilities*, 2012.

[9] M. Mazzucato, *The value of everything: making and taking in the global economy*, London: Penguin, 2019.

G. Standing, *The Precariat: The new dangerous class*, London: Bloomsbury Publishing Plc, 2011.

twentieth century, researchers and policy-makers were largely open and inclusive about their 'birth.' In the US, the Communications Decency Act (CDA) and the Digital Millennium Copyright Act (DCMA) were passed in 1996 and 1998, respectively. The famous Section 230 of the CDA provides a broader exemption for digital platforms; that is, they will not be liable for law violations of the law by third-party subjects on their platforms, except for copyright infringement. Similarly, the DCMA goes further by including copyright liability in its scope of protection, providing that digital platforms are not jointly and severally liable as long as they do not know about of the infringement, do not directly benefit from the infringement and have a 'notice-and-delete' post-facto mechanism. Thus, the CDA and DCMA established the 'safe harbor principle' for digital platforms at their inception. According to Georgetown University Law School Professor Anupam Chander, the law's inclusive rules ultimately 'made the rise of Silicon Valley possible.'[10]

The reasons why the legal rules at the time granted such a broad exemption to digital platforms came from two main sources: on the one hand, in order to protect innovation and promote the development of new industries, the digital platforms had not yet grown into the 'behemoths' they are today and their relatively weak position was not sufficient to trigger systemic governance risks; on the other hand, perhaps more importantly, this was due to concerns that the 'excessive censorship' of platforms might influence political opinion. The rise of digital platforms has changed the structure of public opinion in the 'mass media,' providing a vehicle for the decentralized dissemination of information while dismantling the centralized power of a few institutions to control the channels of communication. In the face of the vast amount of content on the platforms, digital platforms are unable to censor all content legally, similar to traditional mass media, so if they are held jointly and severally liable for illegal content, they may 'over-censor' and exclude legal content from political opinion, thus affecting the overall progress of freedom of expression. In this sense, policy-makers, particularly in the US, have created a 'safe harbor' for digital platforms more broadly.

The 'safe harbor' for digital platforms goes beyond the exclusion of joint and several liabilities, as the technological neutrality and even the 'right to freedom of expression' of digital platforms themselves have been

[10] Anupam Chander, How Law Made Silicon Valley, *Emory Law Journal*, *63* (2013), p. 639.

gradually established in the course of subsequent historical evolution, thus giving digital platforms more rights and less responsibility. Dr. Zuo Yilu of Peking University Law School provides a more detailed account of this historical evolution in an extended discussion in Chapter 5 of Beyond the 'Street Corner Speaker': The Margins and the Centre of the Right to Expression. In the 2003 case Search King v. Google, the US District Court for the District of Oklahoma recognized that Google's search ranking algorithm should be considered an 'expression of opinion' and therefore equally protected by the right to freedom of expression; that is, Search King could not require Google to include it in the ranking of web pages or the position in which it was ranked. In the 2006 case of Landon v. Google, the US District Court for the District of Delaware similarly upheld the claim that Google's algorithm is a freedom of expression, which includes both the 'freedom to say' and the 'freedom not to say,' and therefore, Landon could not require Google to present its own website ads or to display its own website advertisements or to rank them in a particular position. At the same time, the District Court of Delaware further extended the interpretation of CDA 230, holding that even though Google's algorithm was deemed to be protected by freedom of expression, it was not jointly and severally liable because of the immunity doctrine of CDA 230. In other words, Google enjoyed the rights of a speaker without the obligations of a speaker.

As a result of these changes and the establishment of the legal system, digital platforms have been largely insulated from political debate and given free and open space to develop under the broad 'safe harbor principle,' and they have grown rapidly in this market environment. However, with the Cambridge Analytica incident as a sign that countries worldwide are beginning to realize that the scale and influence of digital platforms are no longer the same, a change in regulatory philosophy and legal regime is inevitable. The United States, the first proponent of the 'safe harbor' principle for digital platforms, is experiencing such changes.

In a 2018 interview, US Congressman Chris Cox, one of the authors of the CDA's Section 230, suggested that the content of Section 230 should be reworked. For Cox, the core purpose of Section 230 is 'to help clean up the Internet, not to facilitate people to do bad things.'[11]

[11] Alina Selyukh, Section 230: A Key Legal Shield For Facebook, Google Is About To Change, https://www.npr.org/sections/alltechconsidered/2018/03/21/591622450/section-230-a-key-legal-shield-for-facebook-google-is-about-to-change.

Under the pre-Section 230 legal system, if digital platforms managed the content posted by users, they could be considered 'Publishers' and therefore liable, and this would inevitably lead to digital platforms 'allowing' the proliferation of content on them. The logic of Section 230 is that a broad exemption from liability for digital platforms would remove the fear of being a 'publisher' and allow them to proactively explore content management measures to protect their reputation, thus contributing to creating a clear cyberspace. However, the reality is complicated by the fact that 'reputation mechanisms' have not played an important role in developing digital platforms; on the contrary, they have even benefited from the richness and diversity of content. To a certain extent, harmful content that does not conform to social norms (such as false content or hate speech) serves the commercial interests of digital platforms. Because of this changing reality, the call for an amendment to Section 230 has become a focal point in the US, with the current US President, Joe Biden, even suggesting during his campaign that Section 230 should be 'repealed immediately.'

Exactly how this will change in the future remains to be seen, as does the expansion of digital platforms' size, which increases their bargaining power in the legislative revision process. However, regardless of how Section 230 reform will proceed, it is important to note that how digital platforms manage their content and the manipulation of political opinion that can result from poor management is only one type of political risk that digital platforms can pose, and other political risks of equal or greater importance include the impact and change of traditional political concepts as digital platforms become a universal mode of organization for society as a whole. The impact and transformation of traditional political ideas as digital platforms become a universal mode of organization for society as a whole. In his 2014 book To Save Everything, Click Here: The Folly of Technological Solutionism, Belarusian author Evgeny Morozov describes the social changes driven by digital platforms as the rise of 'technological solutionism' and argues that it is replacing traditional 'value-oriented' political processes with functionalist principles of 'efficiency above all.' This will undoubtedly lead to a chain of political risks and changes. In a 2013 interview with The Guardian, Morozov gave the example of how, in an attempt to boost civic turnout, engineers in Silicon Valley might incentivize people to go to the polls by handing out points. However, this approach obscures whether people are motivated to vote for the sake of points or for reasons of political value and participation. Similarly, picking up litter or turning off lights can earn points, but this can insidiously dilute the ethical and emotional motivations for caring about the environment.[12] In fact, the simplistic treatment of technological answers may not only miss the complexities of the real world but also fail to capture the motivations and intentions behind the actions, which are at the heart of traditional political processes. In this sense, the political risks arising from the digital transformation process driven by digital platforms may still not be fully apparent. The risks to a political opinion that we are currently concerned about may only be the 'tip of the iceberg.' The full picture may only be seen when digital platforms replace or transform political processes in all aspects of human society with 'digital solutions.'

Exploration and Future of Global Governance on Transnational Digital Platforms

The natural expansiveness of digital platforms makes global governance closely linked to them. However, in contrast to the traditional issue of global governance of multinational corporations, the transformative nature of the organizational model of digital platforms also determines the specificity of global governance of international digital platforms. Thanks to the interconnected global Internet, digital platform companies are able to reach global users at a low cost in a short period, and their own need for data accumulation and even monopoly is constantly driving the pace of global expansion of digital platform companies. At the same time, considering the complexity and multifaceted nature of human beings and human society, digital platform companies are accelerating their expansion not only in terms of geographical scale but also in terms of business areas in order to collect data on the digital transformation of human society comprehensively and richly, and ultimately to complete the 'technological answers' of the 'ideal' picture.

The above digital transformation and change processes will inevitably clash with human society's traditional values and behavioral paradigms and trigger a backlash or debugging of established governance systems and mechanisms in various countries and fields. The global governance

[12] Ian Tucker, Interview: Evgeny Morozov: 'We are abandoning all the checks and balances', https://www.theguardian.com/technology/2013/mar/09/evgeny-morozov-technology-solutionism-interview.

issues such as cross-border data flows, digital taxes and digital currencies that are the focus of this book are, in fact, specific risks arising from the development and evolution of transnational digital platforms. As the root cause of these governance issues, digital platforms themselves naturally require special attention and focus. Although we have yet to see a global institution or network dedicated to the governance of international digital platforms, the decentralized exploration of different perspectives by stakeholders has led to a number of promising global reform paths from the bottom up. Specifically, these can be summarized as exploring diverse Technology Routes, restructuring shared interests and building binding institutional frameworks.

Firstly, the Diversity Technology Route explores the fundamental rebalancing of monopolistic digital platforms and the incentive structure of 'data silos' by transforming technology solutions to drive digital platforms to compete in the marketplace with high-quality innovation and services, rather than relying on data-based business practice in order to fundamentally addressing governance risks such as monopolistic market structures and privacy violations. The analysis in this chapter has shown that the key to the 'privatization of the social commons' by digital platforms is the creation of 'data shafts.' In order to break the 'data silo,' policy-makers can certainly require digital platform companies to open up their data through the legal system. The EU's General Data Protection Regulation (GDPR), for example, has become a model for data governance by introducing the Right to Data Portability. While returning control of data to the user, it also helps to break the data monopoly of digital platforms, providing more market opportunities for SMEs and thus promoting market competition. However, the introduction of 'data portability' has only limited how digital platforms can control data and has yet to fundamentally change the structure of their data-dependent, data-monopolizing incentives. In practice, for example, data portability is often reflected in the standardization and harmonization of application programming interfaces (APIs) between digital platform companies. In contrast, data remains 'trapped' within and controlled by the platform. Berners-Lee's reflections on this issue, and the governance reforms he has initiated at the technical level, may be more instructive. In Berners-Lee's view, the World Wide Web's failure to separate web pages from data was a design flaw that led to the accumulation of data by digital platform companies and their ability to commercialize it through models such as advertising, which in turn provided further incentives for platforms to collect more user data. In order to change this incentive structure, Berners-Lee initiated the Solid open-source technology solution, which stores data generated by users on the Internet in a specific location and thus separates it from the application (or web page) that provides the service and digital platform companies must first obtain the user's consent to access the data in order to provide the service. The impact of the Solid solution is twofold: on the one hand, the 'data silo' phenomenon is eliminated, as no digital platform company owns user data, and different platform companies can access the same set of user data with the user's consent, thus breaking the network effect and promoting competition in the market; on the other hand, the incentive mismatch in the commercialization of the Internet is eliminated, as the digital platform company's. The motivation of digital platform companies is no longer to collect and control user data but to improve and innovate their products and services, and it is expected that privacy violations, fake news and other Internet chaos will be fundamentally solved. Blockchain and Web 3.0 are typical examples of other Technology Routes that are different from the current digital platform technology solutions, and together, they form an integral part of the search for an alternative 'digital future' in the digital technology world. It does not matter which technology solutions will become mainstream, but whether we can maintain this openness and innovation.

Secondly, the adjustment of the shared benefit structure refers to the 'pyramid' structure of the business ecology built by digital platforms and the formation of an open, reasonable and democratic model for the distribution of platform revenue in order to achieve the universal sharing of the benefits of platform economic development while taking into account the interests of the platform. The source of value creation of the platform economy is mainly based on three factors: the capital and innovative resources invested by digital platform companies to build and maintain the platform, the products or services produced or provided by the platform participants based on the platform and the resources of the social commons accumulated as the scale of the platform participants gradually increases. While the input of digital platform companies must be respected, this does not mean that they can unilaterally determine the percentage or licensing fees; otherwise, there is a negative impact and trend toward an increasingly 'rent-seeking' economy. The latter two factors should also have a place in the distribution of value in the platform economy, and this is an important safeguard and prerequisite for the sustainable development of the platform as an 'ecology,' and therefore in the long-term interests of the digital platform companies themselves. Given the broad, complex and dynamic nature of the platform economy, regulators cannot, in fact, interfere too much with the specific price levels or price structures set by platforms, and the adoption of procedural regulations to adjust and realize the restructuring of shared benefits may be an experimental direction of reform emerging globally. In February 2021, for example, the Australian Parliament passed legislation requiring digital platforms such as social media and search engines to negotiate a revenue sharing with local news content producers to facilitate the latter's sharing in the substantial advertising revenue generated by digital platforms. The underlying philosophy of the legislation is that the advertising revenue generated by digital platforms is not based solely on the platform's products but also on the rich content shared by the platform's users or the relevant pages crawled by the algorithm, all of which are attributed to the hard work of local news content producers. While some digital platform companies have initiated programs such as user sharing schemes, they have yet to result in general, mandatory industry norms and regulatory requirements, and such sharing schemes remain a unilateral decision by digital platform companies without an open and reasoned consultation process. The bill, drafted by the Australian Competition and Consumer Association, attempts to explore this area. While the bill's details are still somewhat controversial, Australia's regulatory exploration has already had an exemplary effect, with the chair of the UK Competition and Markets Commission, for example, indicating that he will be closely monitoring the scheme's effectiveness; similarly, in July 2021, the All-China Federation of Trade Unions (ACFTU) issued the Opinions on Effectively Safeguarding the Labor Rights and Interests of Workers in New Employment Patterns, which explicitly proposed to negotiate on 'piece rate, draw ratio and labor quotas in the industry.' As a result of this document, in November 2021, Jingdong Group held its first collective bargaining meeting comprising representatives of Jingdong Group, Jingdong Logistics employees and representatives of platform companies. The progress of the relevant reforms remains noteworthy. However, they all undoubtedly reflect the basic idea of moving toward establishing an open, reasonable and democratic model for the distribution of platform earnings.

Finally, the development of binding institutional frameworks remains a stakeholder endeavor. Although the complexity of digital platform structures and behaviors hinders the implementation of specific regulatory measures, and the lack of information asymmetry or expertise still makes regulatory lags a central issue in the global governance of digital platforms, this does not mean that it is impossible to draw 'red lines' for the behavioral strategies of digital platforms. Global governance processes around cross-border data flows, digital taxation and digital currencies are all coming up with binding global institutional rules to influence the behavior and strategies of digital platforms. Robert Fay, a researcher at the Centre for International Governance Innovation (CIGI), an independent Canadian think tank, has suggested that the Financial Stability Board (FSB) should be modeled on the one set up by the G20. The Digital Stability Board (DSB) should be established as the main platform for dialogue on the global governance of international digital platforms, modeled on the Financial Stability Board (FSB) of the G20. However, there are still many challenges and obstacles to this effort. The breadth of issues covered by global digital platforms, the different levels of governance needs and development in different countries and the uncertainty of what future governance reforms might bring make it difficult to establish 'red lines' for global digital platforms in a short time. For this reason, it may be imperative to first reach a consensus on certain aspects of governance concepts to advance the global governance of international digital platforms. A hopeful conceptual consensus might be to continue the policy discussion on 'Network Neutrality' and re-establish the regulatory principle of 'Platform Neutrality-governance consensus.' Net neutrality is a policy option that was discussed at the beginning of the commercialization of the Internet, with the main concern being the right of network operators providing network services to digital business platforms to charge differentiated prices for different forms of data traffic. In order to facilitate the development of new forms of business and maintain the Internet's openness, countries eventually adopted, to varying degrees, the principle of 'net neutrality,' i.e., requiring network operators to treat different data traffic equally. Similar to the network operators of the time, the current digital platforms also have a responsibility to promote ecological development and keep the Internet open. The EU's antitrust investigations into Google and Facebook have focused on whether digital platforms have violated the principle of 'neutrality' by giving more relative advantages to ecological players associated with them. In February 2021, the Anti-Monopoly Commission of the State Council issued its 'Anti-Monopoly Guidelines on the Platform Economy,' which also covers the principle of 'neutrality' of platforms in Articles 12 to 16. Although there is still much controversy surrounding the meaning and validity of the principle of 'net neutrality,' this should be seen as room for future reform rather than a reason to deny the need 'net neutrality.'

Although the governance risks posed by digital platforms are often domestic rather than global (except for digital taxation issues arising from large-scale tax avoidance on digital platforms), the similarity of the relevant governance risks makes the governance practices of different countries extremely informative and useful. In this sense, the global governance of international digital platforms remains indispensable. The three areas of reform explored above undoubtedly all have the potential and need to do so. However, there is still uncertainty about where future reforms will lead. The only certain thing is that the interaction between technological innovation, business development and governance construction will be closer and more profound than ever before, and that the evolutionary changes in the environment built by this interaction will come more rapidly on digital platforms.

PART II

Path to Reforming Platform Governance in Chinese Context

CHAPTER 6

Technology, Governance and Politics the Ideological Foundation of Platform Governance in China

While the discourse of 'science and technology' has persisted throughout the socialist construction phase of the People's Republic of China and its reform and opening up, its policy status and theoretical significance have rarely been studied. One popular understanding is provided by the American sinologist D. W. Kwok, who argues that although conditions for the full application of science in China were scarce in the first half of the twentieth century, this stimulated extraordinary intellectual esteem for science, which became the path to national prosperity in the minds of the Chinese, creating a kind of 'scientism-only' teaching—'The dogma of Scientism.'[1] There is considerable evidence that seems to support D. W. Kwok's view, such as the military and mechanical production practices of the foreigner scholars in the late Qing Dynasty, Yan Fu's focus on the promotion of modern science as a value system in his translations, the May Fourth intellectuals' eulogy of science and democracy and industrial society, and the large-scale appointment of professional intellectuals in different periods of history after the founding of New China and so on. In the words of Hu Shih in 1923: 'During these thirty years, one term has achieved a position of almost supreme dignity in China; neither

[1] Kwok, D. W., *Scientism in Modern Chinese Thought (1900–1950)*, Nanjing: Jiangsu People's Publishing House, 1998.

Y. Jiang and K. Jia, *Transformation of Platform Governance in China*, IPP Studies in the Frontiers of China's Public Policy,
https://doi.org/10.1007/978-981-99-6456-7_6

those who know it nor those who do not know it, neither those who are old-fashioned nor those who are new-fashioned, dare to express their contempt for it openly or to tease it. That term was 'science.' Another question is whether such an almost unanimous reverence for it has any value at all. We can at least say that since China spoke of the Reformation and the Restoration, no self-appointed new figure has dared to slander 'science' openly.[2] But looking back at Chinese political practice from the early twentieth century to the present, 'science' has never achieved such a high status, not to mention the dogmatic notion of 'scientism alone' that has developed in the minds of the nation. The concise account in this chapter attempts to reveal an explanation that is closer to the actual historical process: scientism did not become an 'ideology' in the construction of the nation-state under the leadership of the CCP. However, it was always subordinated to the adjustment of the party line and the general policy. This situation was dictated by the fact that the CCP was implementing nation-state building and social reconstruction at a time when Chinese society was in a crisis of total collapse.

Scientism Revisited

The period around May Fourth was the most active historical period for discourse on science. Although influential scholars such as Joseph Needham (1900–1995) argued for the ancient China's brilliant achievements in science and technology, the only time in Chinese history when science was discussed extensively in the public sphere was around the May Fourth Movement. At that time, science and democracy went hand in hand and were promoted by their advocates as the cure for China's accumulated weakness and poverty. Chen Duxiu, the founder and early main leader of the Chinese Communist Party, was one of the most active intellectuals advocating 'science' in China's intellectual and cultural circles, and in the tone of a battle cry, he proposed that:

> Those who oppose the New Youth are just because we have destroyed Confucianism, rites and laws, national essence, chastity, old ethics, old art, old religion, old literature, and old politics, to name a few crimes. We confess these crimes. However, we have committed these heinous crimes only because of our support for Messrs Nad Mokrasi and Science. To

[2] Quoted in Kwok, D. W., *Scientism in Modern Chinese Thought (1900–1950)*, pp. 8–9.

> uphold Mr. Nader, one had to oppose Confucianism, ritual, chastity, the old ethics, and the old politics. To support Mr. Nasser, one had to oppose the old arts, the old religion. If one wants to uphold Mr. De(democracy) and Mr. Sai(science), one has to oppose the national essence and the old literature. Only after the Westerners had made so much trouble and shed so much blood because of their support for Mr. De and Mr. Sai, Mr. De and Mr. Sai gradually rescued them from the darkness and led them into the light. We now believe these two gentlemen are the only ones who can save China from all the political, moral, academic and intellectual darkness. If we support these two gentlemen, we will not resist all the government's oppression, all the attacks and curses of society, or even the shedding of blood on our heads.[3]

The juxtaposition of science and democracy has shown that the advocacy of science in the *May* Fourth period was not aimed at pursuing pure knowledge but carried a distinctly political stance and even had a compulsive and planning progressivist ideology as its basis. Chen Duxiu equated science and democracy with the well-traveled path to the light while categorizing routes different from their initiatives as authoritarian and superstitious. He said: 'There are now two paths in the world: one is the light path towards the republican science of godlessness; the other is the dark path towards the despotic superstition of theocracy. If the Chinese people want the Boxer Rebellion to cease and hate the idea that such shameful monuments as the Clintons will never be erected again, which is the best path to take?'[4] This clear-cut political stance was not so much related to the communist philosophy that had just been introduced into Chinese intellectual and cultural circles at the time but was inherited from the reaction of the Chinese scholars in the modern era when they were confronted with the all-round challenges of Western technology, institutions and culture, such as the criticism of the Westerners against the reformists in the late Qing Dynasty, i.e., the dichotomy of progressive versus conservative, modern versus retro, legal versus ritualistic, which marked the Chinese nation. The dichotomy of progressive versus conservative, modern versus archaic and legal versus ritual identifies the different

[3] Chen, Duxiu, Reply to the Crime of the New Youth, *New Youth*, 6(1), January 15, 1919.

[4] Chen, Duxiu, The Clintide Tablet, *New Youth*, 5(5), October 15, 1918.

options available to the Chinese nation in the face of the major crisis of foreign invasion and internal collapse.

However, it is also because of this overly political (and consequently compulsive and planning) nature that the promotion of science and democracy by Chinese intellectuals such as[5] Chen Duxiu has drawn criticism from some intellectuals who were not even opposed to scientific and democratic ideas. Lu Xun, for example, criticized the nation in a difficult-to-read literary text a few years before the May Fourth Movement for seeing the democratic constitutionalism ('pluralism') and the scientific spirit ('materialism') transmitted from the West as the basis for national salvation, criticizing respectively: 'Therefore The majority of the people are friends, and the paths of right and wrong are confused; only common words are understood, and they are indifferent to the mysteries.' 'If the matter is exhausted in material things, what is the essence of life in material things? In the ordinary sense of the word, it must be otherwise.'[6] Translated into the terms of modern political theory, Lu Xun's criticism is that Western democracy cannot solve the problem of 'tyranny of the majority' (because majority opinion does not equal truth and benevolence) and that the scientific method cannot solve the problem of the meaning and value of life (because of the dichotomy between fact and value revealed by David Hume).

However, both advocates and critics are discussing science and democracy in the context of the common historical background of the Chinese nation's struggle for salvation and survival, and it is only by understanding this that we can understand the ideology of the construction of the Chinese nation-state and the place of science in the Chinese political process. In fact, in the social and cultural context of modern times, there are very few intellectuals who have explored issues on a purely intellectual level, whether it be the 'Science and Genius Controversy,' or the New Culture Movement itself, or the great debates on the nature of Chinese social history and so on. In these large-scale social and cultural discussions, the participants were all trying to find a way out for the Chinese

[5] In Chinese intellectual circles around the May Fourth period, the great polemic against scientism was the so-called metaphysical school, also known as the polemic of the outlook on life. For a detailed discussion, see Wang, Hui, *The Emergence of Modern Chinese Thought*, vol. 2, Peking: Sanlian Bookstore, 2008.

[6] Lu, Xun, On Cultural Bias to, in *The Complete Works of Lu, Xun*, vol. 1, Peking: People's Literature Publishing House, 2005.

nation. In all these large-scale social and cultural discussions, the participants were sought a way out for the Chinese nation. It is, therefore, understandable why Hu Shih, who also advocated science and democracy, introduced the pragmatism of his teacher, the American philosopher John Dewey, into China, and immediately became the most influential ideological approach of the time. As Qu Qiubai later reflected, 'It was no accident that Hu Shih's experimentalism emerged during the May Fourth Movement in China. Chinese patriarchal society, shaken by the encroachment of international capitalism, demanded a new cosmology and a new outlook on life in order to adapt to the new environment in which China found itself, and the philosophy of experimentalism was just able to meet this need with its positive aspects.'[7]

Qu Qiubai criticized the scientific outlook of the pragmatic philosophy propagated by Hu Shih precisely from the perspective of not completely transforming the existing system and rejecting revolution:

> A country as culturally backward as China, in between the international competition, certainly needed scientific knowledge to cope with it, so experimentalism brought the scientific method to China. This is a historical misunderstanding. Experimentalism can only recognize some practical scientific knowledge and methods not scientific truth. The very character of experimentalism lies in denying the absolute value of all theories. It is the practical philosophy of European capitalist society, especially 'Americanism.' Experimentalism strives to synthesize the psychology of the modern mercenary and implicitly construct a system, although he himself denies all philosophical systems. What the mercenary needs is 'a little of this and a little of that': a little science, a little religion, a little morality, a little worldliness, a little technical knowledge, all of them, but none of them complete so that he can get on with the world. However, one must only go through part of the process, otherwise the conclusion will be that the existing system must be completely transformed, and that this must be done by revolutionary means. How terrible that would be! The status quo can be transformed, but it is not necessary to radically change the

[7] Qu, Qiubai, Experimentalism and Revolutionary Philosophy, in *Selected Writings of Qu, Qiubai*, Chengdu: Sichuan Literature and Art Publishing House, 2010. Qu's use of the term 'experimentalism' is perhaps a better translation of pragmatism than the usual translation of pragmatism as 'pragmatism' by Chinese scholars. I am grateful to Professor Zhiyuan Cui of Tsinghua University for pointing this out.

> existing system, to deal with it in a piecemeal manner. So experimentalism is pluralism, it is reformism.[8]

It is paradoxical, however, that even if one does not accept the 'vulgar' philosophy of pragmatism, as in the case of revolutionaries who put more emphasis on 'daring the sun and the moon to change into a new sky,' one still sees science from a pragmatic standpoint—science is a mechanism for achieving a political goal, which in itself hardly constitutes an ideological basis. Science is a mechanism for promoting a political goal and hardly in itself constitutes the ideological basis for that goal. One can compare some examples of the treatment of science, education and intellectuals in different periods of Chinese history: in the political climate of the 1950s, a very important slogan was popular: 'Education must serve proletarian politics and must be combined with productive labor,' and for this reason, the study of scientific knowledge in Chinese universities gave way to training in. The tutorials of engineering universities such as Tsinghua University were modified to include instructions for students to participate in the design and construction of large reservoirs and other public works projects, and classroom education was replaced by regular participation in manual labor by teachers and students, with the goal of university education changing to the training of 'socialist conscious and educated laborers' as instructed by Mao Zedong. In the 1960s, the party organizations in factories consisted of an overwhelming majority of workers. It was much easier for workers to join the party than for technical cadres, but since the 1980s, an increasing proportion of technical cadres have joined the party. The main management responsibilities in factories have gradually shifted to administrative and technical offices, rather than to the frontline of the shop floor. By the time of the reform and opening up, according to a survey of urban Chinese by sociologist Andrew Walder, the presence or absence of a university degree was not an important requirement for Party membership in Mao's time. In contrast, after 1987, university graduates were nearly six times more likely to join

[8] Qu, Qiubai, Experimentalism and Revolutionary Philosophy, in *Selected Writings of Qu Qiubai*, Chengdu: Sichuan Literature and Art Publishing House, 2010.

the *Party* than non-university students.[9] These examples show that scientific and technological competence, as marked by educational attainment, did not have the same status in the political system in different historical contexts, but that the fundamental political-scientific relationship has remained unchanged and that it is still the political leaders' vision (political philosophy) of the path of revolution and construction of the Chinese Communist Party and the future of Chinese society that plays a decisive role: the dominant political line in the 1950s and 1960s was. After the 1980s, it was a period of economic construction with science and technology as the first productive force, and knowledgeable and educated university students became the mainstay of all walks of life.

Mao's theory of 'combining the principles of Marxism with the practice of the Chinese revolution' is a profound expression of the understanding of the relationship between science and politics:

> To bring the theory of Marxism-Leninism and the actual movement of the Chinese revolution together is to go to it to find positions, perspectives, and methods for solving the theoretical and tactical problems of the Chinese revolution. This attitude is the attitude of having a target. The 'of' is the Chinese revolution, and the 'vector' is Marxism-Leninism. That is why we Chinese communists are looking for this 'vector,' to shoot at the 'the' of the Chinese Revolution and the Eastern Revolution. This attitude is the attitude of seeking truth from facts.[10]

No scientific knowledge or theoretical system is a sacred object that cannot be separated and must be worshipped, but rather a toolbox from which methods ('vectors') can be found to deal with practical problems ('the'). In his speech at the opening of the National Conference on Science on 18 March 1978, Deng Xiaoping established the political status of intellectuals as 'part of the working class itself' and stressed the pivotal role of science and technology in the building of the Party and the state ('science and technology are called productive forces, and scientists are laborers'). Science and technology are the workers, and Deng Xiaoping's more famous assertion in 1988 that 'science and technology are the

[9] The above example, taken from Zhou, An (Joel Andreas), *The Rise of the Red Engineers: Tsinghua University and the Rise of China's Technocratic Class*, Hong Kong: Chinese University of Hong Kong Press, 2017.

[10] Mao, Zedong, Transforming Our Learning, in *Selected Works of Mao, Zedong*, vol. 3, Peking: People's Publishing House, 1991, p. 801.

first productive force' demonstrates this fundamental relationship between science and politics. The reason for this difference from previous historical periods is that the Party's line and approach have changed significantly with the times. The political scientist Tang Zou proposed the concept of the Middle Course to describe the change in the line of the Chinese Communist Party during the Deng Xiaoping era, namely the distinction between the 'economic construction as the center' and the previous 'class struggle as the platform' and the 'adherence to the four basic principles' and the political liberalization. Zou honestly stressed that no matter how wide or narrow the boundary of this 'middle way' was; it was only natural that the most important task was to maintain and uphold the Party's leadership position.[11] The relationship between science and politics, as demonstrated in the revolutionary and founding practices of the Chinese Communist Party, is so profound that it can generate new inspirations and thoughts on institutional innovation at a time when the new generation of the information technology revolution is inspiring great fears worldwide. In short, a review and reflection on the initiatives, debates and practices of science in China in the modern era should recognize the political significance of these, as at no time has 'scientism' become the ideology of China's political and intellectual elites.

In any case, the need to reorient the application of science and technology through political ideas existed in China under the governance of the Communist Party of China, and scientism was not sufficient in itself to become the ideology for the construction of the nation-state under the Communist Party. This state of affairs was determined by the historical fact that the CCP had implemented revolution, nation-building and social reconstruction in a crisis in which Chinese society was in total collapse in the recent past. However, in reviewing the CCP's governance practices, it is also sobering to realize that the CCP's understanding of the political nature of science is a 'double-edged sword' that has the potential to both undermine the constructive function of science and technology (as demonstrated by the repression of elite intellectuals during the Mao era) and, under immediate conditions, adapt the Technology Route that can be adapted to best serve political goals under immediate conditions (as demonstrated in the Maoist era through mass responses to schistosomiasis and the invention of Chinese character picking and

[11] Tang, Zou, *The Cultural Revolution and Post-Mao Reforms: A Historical Perspective*, Chicago and London: The University of Chicago Press, 1986, pp. 219ff.

typography[12]). The key here is: firstly, is it possible to have a clear understanding of the task of party governance of the state and the issues of the times it faces? The following chapter will discuss how the party-governed state has implemented response options to new development opportunities and governance challenges and what the problems are in practice against the backdrop of the rise of new information technologies and the development of platforms.

The Rise of New Information Technology and the Transformation of Governance in China

On 20 April 1994, China achieved a functional connection to the global Internet, ushering in the Internet era. In the decades of development that followed, new information technologies have profoundly changed the lives of Chinese people, becoming an important driver of national economic development and driving changes in the country's governance in various fields while also continuing to achieve iterative upgrades of their own. However, this overall thriving development picture is only sometimes harmonious, harmonious and smooth, but often gives rise to conflicts and conflicts that pose new challenges to the country's governance. The Chinese government's practice of governance change in this process is a good example of the strengths and problems of China's approach to science and technology in the context of political ideology.

In response to the rise of new information technologies, China's political leadership is inevitably faced with a double task: first, to implement effective policies to promote the rapid development of the information technology industry, and second, to control and manage the political and social risks posed by new information technologies.[13] The former involves profound questions about the Party's economic development model, while the latter is closely related to the Party's understanding of China's networked society. Both of these fundamental tasks are emphasized in China under the 'height' of political ideology: promoting the rapid development of the IT industry is one of the specific elements of the

[12] Gross, Miriam, *Farewell to the God of Plague: Chairman Mao's Campaign to Deworm China*, California: The University of California Press, 2016. Mullaney, Thomas S., *The Chinese Typewriter: A History*, Cambridge: The MIT Press, 2017.

[13] Zheng, Yongnian, *Technological Empowerment: The Internet, State, and Society in China*, California: Stanford University Press, 2008, p. 49.

national strategy of 'economic construction as the center' defined during the Deng Xiaoping era, and is even more important. It is also emphasized in the current discourse of regional political competition; controlling the risks of new information technologies is one of the core elements of maintaining the security of the Chinese Communist Party in power.[14] We will argue that this high level of emphasis has enabled the CCP to achieve remarkable success in both promoting IT participation and controlling the risks of new technologies, but that this overemphasis has also resulted in an over-reliance on top-down 'command-and-control' mechanisms of governance, leading to short-lived decision-making, which in turn has led to a politicization of the 'Technology Route.' The 'politics of the Technology Route' is seriously under-appreciated.

The Chinese government has mainly adopted a combination of planning and industrial policy support to promote the development of the information industry. The Chinese government's 'planning hand' has attracted international academic attention in recent years, with some Western scholars suggesting that the policy implementation mechanism of planning differs from both the American and European market economies and the former Soviet Union's planned economy, but is a resource allocation mechanism that combines a certain degree of directionality and flexibility, thus avoiding the drawbacks of a market economy that lacks long-term goal orientation and a planned economy that suppresses the market.[15] A few years ago, an important debate arose between two prominent Chinese economists, Justin Yifu Lin and Weiying Zhang, about industrial policy. In fact, for an industry such as the new information technology, which requires a large amount of initial investment in infrastructure construction, it isn't easy to rely entirely on the market to deploy resources. However, it has to be implemented through the state treasury under the guidance of an industrial development plan. The history of the development of China's information industry illustrates the irreplaceable role of industrial policy. However, in order to avoid the formation of oligopolies, duplication of construction and massive waste, industrial

[14] See the article on the prevention and resolution of major risks published in the People's Daily, Playing a good first move, playing a good active battle—Overview of General Secretary Xi Jinping's important discussion on the prevention and resolution of major risks, *People's Daily*, 15 April 2021.

[15] Gang, Phan, China's Wen Delivers Key Targets of Its Development, *International Business Time*, Archived from the original on 19 March 2011, Retrieved 5 March 2011.

policy cannot replace the role of the market and regulatory mechanisms such as anti-monopoly laws and budget laws, and the sensitivity of policies to respond to the market and social changes must be strengthened.

In April 1997, the National Informatization Work Conference adopted the 'Ninth Five-Year Plan for National Informatization and Visionary Goals for 2000,' which included the Chinese Internet in the construction of the national information infrastructure, thus ensuring the status of new information technology as a major core project of the national informatization project. In January 1999, the 'Government Internet Project' was launched, setting off a boom in the construction of government websites. In October of the same year, the Central Committee of the Communist Party of China issued the 'Proposal for the Tenth Five-Year Plan for National Economic and Social Development,' proposing to promote the informatization of the national economy and society vigorously and to drive industrialization with informatization. From 2004 onwards, Chinese Internet companies have been on the second wave of overseas listings since 2000, such as Baidu's listing on NASDAQ in August 2005, which set a record for the highest initial listing day gain since the Internet bubble in 2000. In June 2007, the Eleventh Five-Year Plan for E-Commerce Development was released, establishing for the first time at the national policy level the strategy and tasks for developing e-commerce. In January 2010, the State Council executive meeting decided to speed up the integration of the telecommunications network, radio and television network and the Internet. In February 2012, the Ministry of Industry and Information Technology issued the 'Twelfth Five-Year Plan for the Development of Internet of Things,' emphasizing that in July 2015, the State Council issued the 'Guidance on Actively Promoting the 'Internet+' Action,' which proposed to promote 'the integration of the Internet with all areas of the economy and society to further deepen.' On December 15, 2016, the State Council issued the 'Thirteenth Five-Year Plan for National Informatization,' which sets out systematic development targets for national informatization. In addition to these industry-wide plans and policy deployments, there are also many industry-specific plans and policies such as 'Internet+' for specific industry sectors, e-government, e-commerce, big data, digital finance and so on. In short, after years of continuous promotion, the information industry has become one of the most important sectors in China's national economic system, and it has also achieved multi-point blossoming and comprehensive development within the information industry.

It is not intended here to quote too much from the achievements made by the development of China's information industry at present. However, just a few sets of examples illustrate the effectiveness of China's Internet infrastructure construction field: in 2015, e-commerce transactions reached 21.79 trillion yuan, jumping to the first place in the world, and the national e-commerce transactions continued to expand to 34.81 trillion yuan in 2019, and it was achieved under the overall economic downturn trend, shows the positive impact of the e-commerce platform trading model in activating the consumer market; in terms of e-government applications, by 2015, a unified and complete national e-government network had basically been formed, a basic information resource sharing system was initially built, and services were continuously extended to the grassroots government, with the level of government openness, online business and government-public interaction significantly improved; by the end of 2018, the total number of Chinese domain names reached 37.928 million, the total number of websites reached 5.23 million, the number of web pages reached 281.6 billion, the number of IPv4 addresses was 338, 924, 544, with 41,079 blocks/32 of IPv6 addresses, the 4G network construction network coverage exceeded 98% of the national population, the total scale of 4G base stations exceeded 3.7 million, maintaining the scale of the world's largest 4G network, and the 5G technology R&D Phase III testing of trials has also been largely completed, steadily advancing the commercialization of 5G. China's information industry has achieved a status at least on a scale that is on par with developed countries and regions such as the US and Europe. Of course, many key areas of core technology still need to achieve independent innovation, and there is still room to improve the integration of information technology with the real economy of the manufacturing industry, and so on, from which China needs to make more efforts to solve.

Regarding risk prevention, China has formed a systematic regulatory system and completed the establishment of rules and regulations. The long-standing problem of a miscellany of regulatory bodies and unclearly defined powers of China's network regulation was not resolved until 27 February 2014, when the Central Government established the Leading Group for Network Security and Informatization. In March 1998, the Ministry of Posts and Telecommunications and the Ministry of Electronics Industry merged to form the Ministry of Information Industry to take over the functions of information and network regulation previously undertaken by the Ministry of Posts and Telecommunications and the Ministry of Electronics Industry, respectively. On 11 March 2008, according to the institutional reform plan announced by the State Council, the Ministry of Industry and Information Technology was formed to guide the promotion of information technology and coordinate the maintenance of national information security. It became the competent authority for the Internet industry. However, at the same time, there are as many as 10 higher-level departments that can directly or indirectly manage websites in China, mainly including the Information Office of the State Council, the Network Management Division of the Ministry of Public Security, the Propaganda Department of the Central Committee of the Communist Party of China, the Press and Publication Department, the Ministry of Culture and so on. The long-standing multi-departmental responsibility for network governance in relevant business areas is divided among multiple departments. However, the distribution of powers and responsibilities among them is not clear enough; for example, in August 1998, the Ministry of Public Security established the Public Information Network Security Supervision Bureau, which is responsible for organizing and implementing the maintenance of computer network security and combating online crimes; in May 2001, the Internet Society of China was established; and in June 2010, the People's Bank of China announced the Measures for the Administration of Non-financial Institution Payment Services, which brought online payments under supervision. The industry association focuses on promoting industry self-regulation and building ethics for industry operations. At the same time, the relevant departments of the Ministry of Public Security and the People's Bank of China mainly implement Internet regulation from the perspective of combating cyber-crimes.

In February 1994, the State Council promulgated the Regulations of the People's Republic of China on the Security Protection of Computer Information Systems. Although China has enacted and implemented many laws and regulations to regulate the development of the Internet, there are still numerous problems. On the one hand, the relevant legislation and supporting systems for Internet regulation in China need to catch up and be of high rank. For example, the Decision of the Standing Committee of the National People's Congress on Safeguarding Internet Security promulgated in 2000 and the Decision of the Standing Committee of the National People's Congress on Strengthening Internet

Protection implemented in 2012 have both played an important role in promoting the rule of law in China's Internet regulation. However, because it was issued in the form of a 'decision,' the content of its provisions still leaves much to be desired in terms of clarity. In terms of legal structure, there are relatively few basic laws and more departmental regulations on Internet regulation, and the fragmentation of Internet legislation has led to many departments rushing to the forefront when it comes to rights and shirking their responsibilities when it comes to obligations and duties.[16] On the other hand, at the level of legislative content, the previous laws and regulations should have emphasized the government's management of the Internet, while paying more attention to the protection of citizens' individual rights. In particular, there is too little room for positive interpretation of so-called sensitive events that occur at a certain time, which leads to a spiral of excessive self-censorship before expressing one's views and ultimately discourages people from expressing their views publicly on many serious issues.

In 2013, the US Snowden incident came to light, and the US unilateral cyberspace policy was widely questioned by the international community, which also made countries start to examine their cyber security issues. After 2014, Internet sovereignty and technology nationalism/nationalism concepts and practices have accelerated globally, such as the EU building security defenses through personal data protection regimes, Russia and Brazil strengthening security reviews, and data localization policies have emerged. In China, the Central Leading Group on Cybersecurity and Informatization, established in 2014, is tasked with implementing top-level design for safeguarding cybersecurity, protecting national interests and promoting the informatization development. However, it is worth pointing out that out of the great importance attached to the role of the Internet in fostering new dynamics of economic development and promoting the transformation and upgrading of various types of enterprises, China's network regulation during this period did not aim at strict regulation. In 2015 alone, the State Council issued 12 intensive policy documents related to promoting the development of the Internet, promoting the concept of deep integration of the Internet with all areas

[16] Zhou, H. H., Building the Rule of Law in Cyberspace, http://news.youth.cn/jsxw/201410/t20141026_5910191.htm,2014-10-26.

of the economy and society nationwide, and forming a policy orientation of 'Internet+' to promote new forms of economic and social development.[17]

On 7 November 2016, the 24th meeting of the Standing Committee of the 12th National People's Congress adopted the Cyber Security Law of the People's Republic of China, which stipulates two major initiatives in Article 8[18] and Article 9,[19] the basic basis for establishing a multi-component organic and collaborative cyber governance structure at the organizational and national legal levels. In March 2018, according to the Program for Deepening the Reform of the Party and State Institutions, the former Central Leading Group for Network Security and Informatization was transformed into the Central Committee for Network Security and Informatization, a deliberative and coordinating body directly under the CCP Central Committee, 'responsible for the top-level design, overall layout, coordination, overall promotion and supervision of the implementation of major work in relevant fields.' This series of measures signifies that China has formed a complete network supervision system of 'top-level design - supervision and law enforcement - industry participation,' which establishes a guarantee to overcome the drawbacks of multiple management and overlapping responsibilities.

However, although the governance framework has been established, China has often failed to respond to the development of new information technologies because, on the one hand, new information technologies are still at a rapid stage of development and, on the other, after all, the governance approach is still mainly based on a top-down 'command-and-control' mechanism. In recent years, there have been numerous cyber

[17] Tencent Research Institute, Twenty Years of Internet Regulation in China, https://www.tisi.org/15948.

[18] Article 8 stipulates that 'The State Internet Information Department is responsible for coordinating network security work and related supervision and management. The competent telecommunication department of the State Council, the public security department and other relevant organs are responsible for network security protection and supervision and management within their respective areas of responsibility in accordance with the provisions of this Law and relevant laws and administrative regulations.'

[19] Article 9 stipulates that 'Network operators carrying out business and service activities must comply with laws and administrative regulations, respect social morality, abide by business ethics, be honest and trustworthy, fulfill their obligations to protect network security, accept the supervision of the government and society, and assume social responsibility.'

security incidents of varying degrees of harm, such as P2P mines, personal information leaks, spam emails and malicious attacks on networks, which have raised new demands for continuous improvement of cyber governance mechanisms, and the Chinese government is often caught in the dilemma of either letting things go or controlling them too strictly. What is needed is an institutional mechanism that can help to constantly track changes in the technological sphere and cyber society and to respond to them in a concerted manner. As the analysis later in this chapter will show, China's governance of new information technologies is overly dependent on the 'control-and-command' mechanism. It fails to clearly understand the politics of the Technology Route, resulting in a governance system that lacks sensitivity and foresight in the face of rapidly changing technology and cyber society.

Command-and-Control Mechanisms and Platform Governance in China

China's 'high' political-ideological emphasis on development and technology issues has led to a reliance on 'command-and-control' mechanisms in governance practice. Paradoxically, the dual political need (political demand) for new information technologies—to promote the development of the information industry and to strictly control the political and social risks of new technologies—coupled with the 'command-and-control' nature of the governance mechanism, has led the Chinese Communist Party(CCP) to regulate new technologies instead. This has led the Chinese Communist Party to fail to fully appreciate the profound connotation of the 'politics of Technology Route' in the field of new technology regulation. Two distinct tendencies have thus emerged in Chinese cyber society that potentially challenge the CCP's governance: on the one hand, the technological and business elites that are close to the development of the country's information industry have been allowed to rapidly accumulate wealth and technological resources through policy impetus, and 'digital empowerment' has evolved in a relatively short time to expand the scale, technological capacity and monopoly of a few platforms. On the other hand, the statist-based network regulation system has reinforced the tendency of self-censorship in the online public sphere and the vulgarization of online culture due to the 'clamor of the crowd.' China's new IT governance system, which relies too heavily on 'command-and-control' mechanisms, has revealed serious shortcomings at this point:

policies have had to jump back and forth between a laissez-faire approach under the discourse of promoting development and a strict control under the discourse of security and order, without political deliberation for the benefit of the larger population and future development.

Let us look at the evolution from 'digital empowerment' to the promotion of platform monopoly and from 'the clamor of the crowd' to the vulgarization of online culture. The Technological Empowerment of the Internet has broken the traditional elitist monopoly on the means of production, business information and technological resources, and grassroots netizens have more opportunities to connect and have more productive resources in an equal and open Internet environment. At the same time, the Internet, as a virtual space, provides a convenient and fast platform for the expression of opinions of netizens, and interpersonal communication in the network has changed from point-to-point to point-to-counter, prompting individual consciousness to gather together easily, breaking through the limitations of time and space. Integrating of new information technology with economic and social development has stimulated the creation of new business models and allowed more people to participate in innovation and entrepreneurship.[20] The rapid development of self-media has also provided more and more convenient platforms for ordinary people to express their voices and show themselves.[21]

However, the online society is not isolated from real-world power relations, and real-world factors such as capital power, regional endowments, political power and even individual talent opportunities still impact the use of new information technologies for development. In online regulation, the reliance on 'command-and-control' mechanisms has resulted in overly harsh regulations that stifle the growth of a rich online culture. The usual criticism of China's online regulation in the literature is that the regulatory regime is too harsh and often fails to protect certain normal expressions.[22] However, such observations are too general and do not sufficiently help us to understand the problem.

[20] Parker, Geoffrey G., Marshall W. Van Alstyne, and Sangeet Paul Choudary, *Platform Revolution: How Networked Markets Are Transforming the Economy—And How to Make Them Work for You*, New York and London: W. W. Norton, 2016.

[21] Gibson, Rachel K., Wainer Lusoli, and Stephen Ward, Online Participation in the UK: Testing a 'Contextualised' Model of Internet Effects, *The British Journal of Politics & International Relations*, no. 4, 2005.

[22] How does China censor the Internet?

Indeed, there are still many problems with specific online regulatory measures, one of the more prominent of which is China's 'principle of strict authorizations' for media services, including the Internet, which is the most stringent in terms of restricting the content of illegal information. For example, not only is information that reveals state secrets, the transmission of pornography to minors, information that promotes violence and racism, and false information that threatens and intimidates others, which is strictly prohibited in many Western countries, but also 'encouraging mobbing,'[23] 'inciting unlawful assembly,' and 'using the Internet for the purpose of gathering people,'[24] 'using the Internet to organize contact with cults,'[25] 'endangering social morality or national cultural traditions,'[26] etc., are not clearly defined or prohibited in most of the major countries of the world today. The biggest problem, however, does not lie in the heavy-handed content regulation described above. Since the beginning of the century, national regulators have developed and applied technical mechanisms such as keyword filtering of IP messages to monitor most of the information on international entry points to the Internet, direct IP address blocking at national entry gateways, keyword blocking on backbone routers and domain name filtering by controlled DNS, to very strictly control the freedom of the Internet, making it a veritable most stringent network regulation system. This is the over-reliance of network regulation on the 'command-and-control' mechanism: (1) over-reliance on automatic detection technology with sensitive word control, traffic control, etc.; (2) over-reliance on the reporting mechanism for Internet rumors, undesirable information, etc. The over-reliance on top-down 'command-and-control' mechanisms leads to the simple and brutal blocking of online information without regard to substance or position, which in the long run breeds a certain cultural mentality of suspicion and mistrust (i.e., the suspicion that behind any unspeakable event there is a hidden 'conspiracy') on the one hand, and a 'conspiracy' on the other. Because it does not lead to targeted arguments and

[23] For example, Article 18(1) of the Provisional Measures for the Administration of Educational Websites and Online Schools.

[24] For example, Article 19 of the Regulations on the Administration of Internet News and Information Services.

[25] For example, Article 2 of the Decision on Safeguarding the Security of the Internet.

[26] For example, Article 14 of the Regulation on the Administration of Business Premises for Internet Services.

rebuttals, it opens the door to disinformation, slander or purely personal expressions, making it difficult to create an orderly, healthy and high-quality discussion on major issues on the Internet, and instead only allows for banter, irony, parody and entertainment. They can only air their grievances through banter, irony, parody and entertainment. The urgent need to rebuild the public culture of the network society due to the 'clamor of the crowd' cannot be achieved under the over-reliance on the top-down 'command-and-control' mechanism.

In addition, over-reliance on the 'command-and-control' mechanism has also led to a lack of intrinsic motivation for the development of the rule of law in network regulation, resulting in China's regulation of responsibility for Internet information dissemination and network security, which is mostly in the form of principles and general statements and lacks operability in practice. Under such strict liability regulation in principle, Internet enterprises impose layers of severe regulations on information posted by users to avoid their legal responsibilities, with online information content being deleted at every turn, and the only remedy available is to appeal to the service provider's platform. This situation exacerbates the trend of fragmentation in the development of the system in the field of Internet regulation in China; that is, it is impossible to promote the development of the Internet legal system through equal consultation and rational communication between the regulatory authorities and the industry, society and Internet users.

On the surface, the problems of online regulation appear to be the consequence of an overly 'political' regulatory practice. However, a closer analysis reveals the opposite: automatic detection technologies and reporting mechanisms have brutally blocked the possibility of rational political deliberation. In fact, before the emergence of new information technologies such as the Internet and social media, the regulation of public opinion in countries around the world was based on political deliberation, i.e., on safeguarding the general political situation, regulating the balance between different kinds of expression, allowing opinions to be expressed from both the left and the right, while at the same time confronting them and preventing extremist tendencies. With the over-reliance on 'technical governance' such as automatic detection technology and reporting mechanisms, the filtering mechanism itself has no 'big picture' to speak of, and reporting can easily be reduced to venting personal anger and denouncing each other. The current regulatory mechanism, which appears to be quite stringent, in fact, has very negative consequences for the maintenance of the overall political situation (as required by the CCP to maintain the security of its rule): for example, (1) the so-called 'high-class black' and 'low-end red' speech is often not effectively intervened, affecting the people's trust in and respect for the Party and the government; (2) Nasty and vulgar information is not regulated promptly, resulting in the deterioration of the Internet environment; (3) legitimate exchanges on serious topics and moderate emotional outbursts are brutally curbed, exacerbating the discontent of ordinary Internet users. Not long ago, the state Internet regulator blocked many public websites that were supposedly in the 'patriotic business' of smearing other countries with rumors under the banner of 'patriotism,' even causing some countries to protest China during the prevention and control of the new epidemic. The persistence of such public websites also shows that the more important responsibility of Internet regulation is to cultivate a rational, thoughtful and discerning attitude among Internet users to promote the formation of a clean public culture in online society.

The literature on the politics of communication shows that virtual organizations formed by the pooling of individual consciousness are very sticky to individuals and will change as individual consciousness is pooled or dispersed, so that in order to gain more followers (the so-called fans or attracting traffic), the 'crowd noise. Therefore, in order to gain more followers (i.e., to 'attract fans' or attract traffic), the 'clamor' of the Internet culture would have formed a tendency of popularization or even vulgarization and vulgarization, which would have destroyed the survival base of traditional and elite culture on a large scale.[27] What is needed to counteract this trend of banalization and vulgarization is to use the advantages of new information technology in reducing the cost of communication and to maintain the 'politeness' of the online public sphere with wider public participation and public discourse.[28] However, because China's 'command-and-control' mechanism of Internet censorship is too strict, many serious public issues cannot be openly discussed, and the increasingly entertaining online orgies are largely beyond the

[27] On the issue of mass media and the reconstruction of publicness under the conditions of modernity in contemporary China, see Lv, Xinyu and Yuezhi Zhao, Modernity, Mass Media and the Reconstruction of Publicness in China, in Love Thought website: http://www.aisixiang.com/data/34285.html.

[28] Rasmussen, Terje, Internet and the Political Public Sphere, *Sociology Compass*, 8(12), December 2014.

scope of regulation, resulting in a very serious tendency toward the banalization and vulgarization of Internet culture.

Similarly, digital resources are not equally empowering for every participant. Rather, due to the multiplier effect of the network, the new information technology has led to faster and larger wealth accumulation, and in the absence of strong regulation, 'digital empowerment' has not narrowed but rather widened the gap between the rich and the poor.[29] The Chinese government's over-reliance on the 'command-and-control' mechanism in promoting the development of the information industry is mainly reflected in its strategies and policies: planning and industrial policies in the field of information technology and related science and technology adopt the mindset of a catch-up strategy, i.e., 'adopting a distorted product and factor price approach and replacing the market with a planning system.' The approach and the institutional arrangement of replacing the market mechanism with a planning system to improve the country's ability to mobilize resources break through the constraints on the development of capital-intensive industries due to the comparative disadvantage of capital scarcity and thus make the industrial structure reach the level of the first developed countries.[30] The implementation of the catch-up strategy inevitably requires a large set of management systems and project operating procedures, and the 'command-and-control' mechanism has a very strong sense of process and management habits. Accordingly, the sensitivity to factors such as the subjective perception of scientific and technological talents engaged in cutting-edge work, market demand and social needs is greatly reduced. For late developing countries, a catch-up strategy can achieve considerable economic development.[31] However, for scientific and technological fields that require more innovative thinking and cutting-edge exploration, a catch-up strategy that relies excessively on a 'command-and-control' mechanism is highly inadequate—and this is exactly what China is

[29] Zuboff, Shoshana, *The Age of Surveillance Capitalism: The Fight for a Human Future at the New Frontier of Power*, Profile Book, 2019.

[30] Justin Yifu Lin, *Interpreting the Chinese Economy*, Peking University Press, 2012, Lecture 4.

[31] Rock, Michael T. and Michael A. Toman, *China's Technological Catch-Up Strategy: Industrial Development, Energy Efficiency, and CO_2 Emissions*, Oxford: Oxford University Press, 2015.

currently doing in the mid-1990s. This is China's biggest dilemma facing in the US-China trade and technology wars.

The development of the Internet in China has so far led to the formation of the so-called three-legged triumvirate, i.e., BAT (Baidu, Alibaba and Tencent), the three largest Internet companies in China, which have developed their systems and strategic plans to hold information-based, transactional and relational data in China, and then use their access to the masses to merge with the up-and-coming innovative companies continuously. The story of these three giants illustrates the problem of wealth disparity exacerbated by the 'digitally empowered' nature of China's online society. At the same time, large technology platforms do not guarantee the dynamism of technological innovation, a point that has been well documented.[32] But here, rather than discussing the rise of large technology platforms in China and their ability to create science and technology, we are going to cite the experience of a more familiar industry—online taxis—that developed earlier as a case study to illustrate the impact of the 'command-and-control' governance on the development of technology due to. The process of change from 'digital empowerment' to increased income disparity due to 'command-and-control' governance mechanisms is somewhat common in networked societies. This is because when the industry first emerged, it was customarily seen as representative of the 'sharing economy,' which was often thought of as 'co-production and sharing' through the Internet. However, the following review shows that this is different.

From 2016 to 2019, the transaction scale of China's online taxi market increased year by year, and in 2019, the overall transaction value of China's online taxi market reached RMB 304.41 billion. The overall transaction amount reached RMB304.41 billion. However, in such a huge market scale, the online taxi market has long formed an oligopoly: in 2015, Drip (which has received multiple rounds of investment from Tencent, CITIC, CIC and Temasek) accounted for 99% of China's online taxi market share, 87% of the online car market and over 70% in other business areas.[33]

[32] Lazonick, William, *Sustainable Prosperity in the New Economy: Business Organization and High-Tech Employment in the United States*, W. E. Upjohn Institute, 2009.

[33] Dai, J., and J. Tong Tian, *A Study on the Regulation of Emerging Internet Industries*, Peking: China Development Press, 2017, pp. 73–74.

The implementation of regulated management at this time has instead facilitated the crowding out of competitors by companies occupying a monopoly position. On 26 *July* 2016, the General Office of the State Council issued the Guidance Opinions on Deepening Reform and Promoting the Healthy Development of the Rental Car Industry, which, for the first time, proposed the inclusion of special Internet cars into the management of booking rental cars and clarified the legal status of net cars; on 27 *July* 2016, the Ministry of Transport, Ministry of Industry and Information Technology and other seven ministries and commissions jointly promulgated the Interim Measures for the Management of Network Booking Rental Car Operation and Services, which established the development positioning and orderly development of online cars according to the principle of high-quality and differentiated operation, and required that localities should adopt local policies and formulate specific implementation rules in conjunction with local realities. Drip responded: starting 1 *January* 2019, Drip removed and stopped dispatching all unlicensed and unqualified vehicles, stipulating that to become a Drip car, you must have two net car certificates, a 'network taxi business license' and a 'network taxi driver certificate.' The fact is that you can find many people who are not able to get a good deal on many things. In response to this situation, some industry analysts have pointed out that 2019 is the year of net taxi compliance and the end of the competition in China's net taxi market.[34]

China's Technological Development: Where to Go?

Our previous analysis has shown that in the Chinese Communist Party's ideology, issues of science and technological development are considered under the party line and guidelines. However, the over-reliance on top-down 'command-and-control' mechanisms has resulted in such 'political deliberation' on science and technology issues not having the effect of opening up multiple paths of development but rather stifling the innovative energy of participants in many areas such as online social regulation and platform governance. In many areas such as network social regulation

[34] Analysis of the size and growth prospects of the online car market in 2020, in Foresight Economist: https://baijiahao.baidu.com/s?id=1667000376138315188&wfr=spider&for=pc.

and platform governance, the innovative energy of the participants needs to be stifled. Is it possible for China's new IT governance to change and to achieve a system of governance conducive to shared freedom? Let us start by looking at the richer connotations of the categories of science and technology and technology governance, reorganize the logic of governance in China and consider new possibilities.

Modern science was born out of the European Enlightenment of the seventeenth and eighteenth centuries. The Enlightenment was a movement of intellectual liberation that dispelled feudal tyranny and religious obscurantism with the light of reason. However, when the 'reason' and 'laws' promoted by the spirit of science were pursued as 'absolute certainties,' it resulted in the suppression of the humanist spirit and the emancipation of individuality that had begun since the Renaissance.[35] In the history of modern European thought, from the philosophy of life of Bergson and Nietzsche, the philosophy of the existence of Kierkegaard and Heidegger, the deconstructionism of Foucault and Derrida, the novels of Dostoevsky and Kafka, the comedies of Chaplin, and even the radical anti-progressivist thinkers such as Miles and Kokos and Carl Schmitt all of them were critical of scientism and the 'technical governance' on which it was based. However, this strong anti-scientist trend often slides toward a complete rejection of the value of science and technology, ultimately confusing ideas and perceptions. For example, the German public jurist Karl Schmitt saw fascism as an overcoming of 'technical governance' such as Anglo-American parliamentary politics in response to the near-destruction of European civilization,[36] the Hungarian philosopher Georg Lukács saw it as itself a manifestation of radical irrationalism,[37] and the sociologists Peter Novick and Lawrence Rees saw it as a radical expression of irrationalism in itself, while the

[35] Toulmin, Stephen, *Cosmopolis: Hidden Agenda of Modernity*, Chicago: University of Chicago Press, 1990.

[36] Schmitt, Carl, The Age of Neutrality and Depoliticisation, in *Schmitt's Collected Works: The Concept of Politics*, translated by Zongkun Liu et al., Shanghai: Shanghai People's Publishing House, 2004.

[37] G. Lukács, *The Destruction of Reason*, translated by Wang Jiuxing, Nanjing: Jiangsu Education Press, 2005.

sociologists Peter Novick and Lawrence Rees revealed it as an extreme expression of rationalism in its neatly planned, mechanistic operation.[38]

Ancient Chinese thought did not have the modern Western 'science' concept. The Chinese philosopher Li Zehou argues that the Chinese cultural tradition is one of the 'pragmatic reasons,' with a preference for practice and utility and a lack of abstraction and normativity, which prevented the development of scientific concepts as pure intellectual inquiry. In contrast to Kant's system of categories, Li Zehou says: 'The greatest flaw and weakness of traditional Chinese pragmatic reason lies in the lack of adequate recognition and full development of the meaning and power of this perceptual abstraction of the nature of practical operation. This is an important reason why China lacks a pure philosophy of highly abstract thought and discernment. As I have pointed out many times before, China has skill, but no science. Chinese culture did not and could not produce Euclidean geometry and Pythagoras because of its overemphasis on "theory in relation to practice" traditional Chinese practical reason placed too much emphasis on the possibility of reality and too little on the possibility of logic, and thus often belittled and devalued "useless" abstract thinking.'[39] After the May Fourth Movement, 'Mr. Sai' came within the purview of the Chinese, but as we had argued before, science and democracy were mainly emphasized as means to help the Chinese nation escape the bullying of the powers and emerge from its destiny of poverty and weakness. The political treatment accorded to science and to intellectuals represented by scientific and technological workers varied from one historical period to another after the Chinese Communist Party gained leadership of Chinese society. However, the fundamental relationship between politics and science remained unchanged: the former's understanding and vision of social development was the decisive factor. The Chinese view of science or scientism remains subordinate to 'practical reason.'

[38] Novick, Peter, *The Holocaust and Collective Memory*, translated by Wang, Zhihua, Nanjing: Yilin Publishing House, 2019; Rees, Lawrence, *Auschwitz: The Nazis& the 'Final Solution'* (*Auschwitz: A New History*), translated by Shuang Liu, Nanjing: Yilin Publishing House, 2020.

[39] Li, Zehou, *Practical Reason and the Culture of Joy*, Peking: Sanlian Bookstore, 2005, p. 12.

This relationship between politics and science in a party-ruled state can only be understood in the relationship between the Chinese Communist Party and Chinese society. In the modern era, the Chinese nation was subjected to a comprehensive military, cultural, economic and institutional onslaught from the West. The pillars of the traditional imperial dynasty collapsed, and the scholarly class, the Confucian moral system, the small peasant economy and the kinship ethic were all treated as negative factors that prevented China from modernizing. Against this historical backdrop of the collapse of social order and the disintegration of human nature, both the Kuomintang and the Communist Party chose to unify all sectors of society through the Soviet Leninist model of party organization, resulting in a political governance pattern that later favored the centralization of power. As this passage from the Constitution of the Communist Party of China, which was partially revised at the 19th National Congress in 2017, shows, with its profound heritage and distinctive contemporary color, the CCP is not external to Chinese society; its self-perception, mission and the way it organizes its power are firmly rooted in Chinese society, which determines that the CCP has the will and ability to make a difference to Chinese society, science, culture, education, economy and so on, This means that the CCP has the will and the ability to lead China in all aspects of society, science, culture, education, economy and so on:

> The leadership of the Communist Party of China is the most essential feature of socialism with Chinese characteristics and the greatest advantage of the socialist system with Chinese characteristics. The Party is the leader of everything in the Party, the government, the military and the people, and in the east, west, north and south. The Party must adapt to the requirements of reform and opening up and socialist modernization, adhere to scientific, democratic and legal governance, and strengthen and improve the Party's leadership. The Party must play a leading and central role in various organizations at the same level, in accordance with the principle of taking the overall picture and coordinating all parties. The Party must concentrate its efforts on leading economic construction, organizing and coordinating the efforts of all parties, working with one heart and mind, and working around economic construction to promote comprehensive economic and social development. The Party must implement democratic and scientific decision-making, formulate and implement correct lines, guidelines and policies, do a good job in organizing the Party and in propaganda and education, and play a pioneering and exemplary role for all Party members. The Party must operate within the confines

> of the Constitution and the law. The Party must ensure that the legislative, judicial, administrative and supervisory organs of the state, economic and cultural organizations and people's organizations work proactively, independently and responsibly, and in a coordinated manner. The Party must strengthen its leadership of trade unions, communist youth leagues, women's federations and other group organizations so that they maintain and enhance their political, advanced and mass character and give full play to their role. The Party must adapt to the development of the situation and changes in circumstances, improve its leadership system, improve its leadership style and enhance its ability to govern. Communist Party members must work closely with the masses outside the Party and work together for the building of socialism with Chinese characteristics.

Of course, it should be immediately clear that it is one thing to understand the place of science within the ideology of the Chinese nation-state and quite another to devise institutional mechanisms to make science work and facilitate the achievement of political ends. In any case, historical developments since the Industrial Revolution in Britain have shown that science and technology have been extremely important and constructive in promoting social progress and economic development and improving the well-being of the people. In this regard, therefore, the significance of Deng Xiaoping's implementation of rectifying the situation and the reaffirmation of the role and place of science and technology in the communist system for China's development cannot be overemphasized. This chapter's criticism of the over-reliance on top-down 'command-and-control' mechanisms is intended to draw attention to the fact that in Communist-governed China, politics can adapt and make effective use of science and technology, which can be fully exploited through changes in the mode and mechanisms of governance. The key question here is what political governance goals should be firmly established, for whose benefit, and with what governance mechanisms make good use of scientific and technological (e.g., new information technology) tools?

Xi Jinping's report to the 19th National Congress has already answered the question of the current political governance objectives of the CCP: firstly, the CCP has adjusted its understanding of the current main contradiction in Chinese society; that is, 'Socialism with Chinese characteristics has entered a new era, and the main contradiction in Chinese society has been transformed into the contradiction between the people's growing need for a better life and unbalanced and insufficient development.' Then, it has proposed the CCP's response; that is, 'We should, based on continuing to promote development, focus on solving the problem of unbalanced and insufficient development and vigorously improve the quality and efficiency of development.' Then, she proposed the CCP's response, namely 'On the basis of continuing to promote development, we should focus on solving the problem of unbalanced and inadequate development, vigorously improve the quality and efficiency of development, better meet the growing needs of the people in the economic, political, cultural, social and ecological fields, and better promote the all-round development of people and the all-round progress of society.' Finally, it emphasized the way forward for the CCP to achieve its current political governance goals, namely 'the Party as a whole should firmly grasp the basic national condition of the primary stage of socialism, firmly base itself on the greatest reality of the primary stage of socialism, firmly adhere to the Party's basic line as the lifeline of the Party and the country and the people's well-being, lead and unite the people of all nationalities, focus on economic construction, and adhere to the four basic principles, adhere to reform and opening up, stand on their own feet, and strive to build China into a rich, strong, democratic, emphasized, harmonious and beautiful socialist modern power.' Through this series of ideological expressions, in short, we can clarify that the task of governance that the CCP has set itself at present is 'to better promote the all-round development of human beings and the all-round progress of society.'

Further, how can the role of science and technology be considered in the context of the governance objective of 'better development of the whole person and progress of the whole society'? We can explore this question at a preliminary theoretical level. A wealth of research shows that science and technology are not neutral and that scientific and technological progress involves the political science of which groups are benefited and in which direction. In Chapter 13 of Volume I of Capital, Marx reveals the socio-political consequences of technological progress, proposing the proposition that 'technology is a social relation unfolding in concrete history,' and the contemporary Western Marxist theorist Jon Elster further analyses this by stating that 'in a given More than one efficient technology in a given time,' thus opening up a variety of Technology Route competition thinking.[40] Empirical studies have shown that there were originally a variety of options for planning technologies such as CNC

[40] Elster, Jon, *Explaining Technical Change: A Case Study in the Philosophy of Science*, Cambridge: Cambridge University Press, 1983, p. 163.

machines, televisions, buildings, urban transport facilities and computer operating systems, some of which were more beneficial to managers, some to workers and some to ordinary consumers.[41] The optionality of the Technology Route creates the conditions for people to participate in technological development and to share in its achievements. The optional nature of the Technology Route creates conditions for people to participate in technological development and share in its achievements. There are many examples of Communist-led nation-building practices that tentatively confirm this view. For example, in the historical context of China's lack of capital and electronic information science and technology, large-scale technological innovation involving the general public had already taken place, such as the 'electronic literacy' campaign in Shanghai and other cities in the 1950s, and the experience of the 'barefoot electricians' among the ordinary workers thus trained in the 1960s and 1970s. In the 1960s and 1970s, the experience of the 'barefoot electricians' who were thus trained among ordinary workers enabled automated production in many industries.[42] At present, when China has reached a considerable level of scientific and technological talent and financial resources, it is all the more important to explore a development model that links science and technology closely with the general workforce, including scientific and technological workers and ordinary workers.

The analysis earlier in this chapter pointed out that in promoting the information industry and regulating the network society, over-reliance on the 'command-and-control' mechanism has led to problems such as oligopolistic markets and the decay of the network public culture. How to change this situation and create a new model of information technology development so that the party-ruled state can effectively promote the development of the information industry and properly regulate the cyber society simultaneously? Before concluding this article, we propose that it is most important to return to the same philosophy that the Chinese Communist Party has never abandoned: 'everything for the masses, everything relying on the masses.' This requires expanding and guaranteeing

[41] Noble, David F., *Forces of Production: A Social History of Industrial Automation*, New Brunswick and London: Transaction Publishers, 2011; Smythe, Dallas, What Comes After the Bicycle?: The Politics and Ideology of Technology, translated by Hongzhe Wang, *Open Times*, No. 4, 2014.

[42] Wang, Hongzhe, From 'Barefoot Electrician' to 'Electronic Bagman': Technology and Labor Politics in China's Electronic Information Industry, *Open Times*, no. 3, 2015.

the participation of industry associations, labor groups and other kinds of social subjects in decision-making under new historical conditions; exploring institutional innovations that make technological governance better serve the political values most cherished by the Communist Party; and making technological governance subservient to democratic political deliberations under the leadership of the Communist Party by strengthening the capacity of the CCP to lead broad democratic participation. Only by firmly implementing the value that technological development is for the benefit of the masses, including workers, can the respective ills of techno-optimism and techno-nationalism, and the tensions between them, be eliminated.

CHAPTER 7

Choice of Technology Routes as a National Responsibility: Change in the Chinese Context?

Introduction: Digital Economy Development Under the Common Wealth Strategy

At the 10th meeting of the Central Finance and Economics Commission held on 17 August 2021, 'achieving more obvious and substantial progress in the common prosperity of all people' was set as one of the visionary goals of socialist modernization by 2035. In his speech, General Secretary Xi Jinping specifically pointed out that 'the new round of scientific and technological revolution and industrial change has given a strong impetus to economic development, and has also brought about profound impacts on employment and income distribution, including some negative impacts, which need to be effectively addressed and resolved.' General Secretary Xi stressed that our common prosperity is 'to prevent social stratification, open up channels of upward mobility, create opportunities for more people to become rich, form a development environment in which everyone can participate, and avoid 'inward rolls' and 'lying flat.'[1] Against the backdrop of the rapid rise of the technological revolution

[1] Xi, Jinping, Solidly promoting common prosperity, Seeking Truth, No. 20, 2021. Available from Seeking Truth: http://www.qstheory.cn/dukan/qs/2021-10/15/c_1127959365.htm. (Last accessed: 10 June 2022, same below).

Y. Jiang and K. Jia, *Transformation of Platform Governance in China*, IPP Studies in the Frontiers of China's Public Policy,
https://doi.org/10.1007/978-981-99-6456-7_7

centered on a new generation of information and communication technologies,[2] the Communist Party of China's Common Wealth Strategy sets out the urgent need to explore institutional and technological conditions that will help the majority of people to participate in development on their own.

Many studies have shown that while most countries have benefited to varying degrees from the process of globalization in recent decades, the uneven distribution of wealth in the world has also deteriorated sharply, and it is no longer simply a case of 'the rich getting richer and the poor getting poorer,' but a complex and serious situation of 'the rich getting richer, the middle-income group getting more divided and the poor getting bigger.' It is a complex and serious situation in which 'the rich are getting richer and the poor are getting poorer.'[3] Many factors have led to the simultaneous growth of wealth and the worsening of inequality, and technological change is one of them that cannot be ignored. In their review of Thomas Piketty's acclaimed book *Capital in the Twenty-First Century* (2014), economists Laura Tyson and Michael Spence criticize Piketty ignored the 'winner-takes-all effect' of information technology development.[4] Robert W. Fogel's econometric study of development data

[2] The following chapter does not strictly distinguish between information technologies, communication technologies and digital technologies but refers to the new generation of information and communication technologies such as mobile Internet, big data, blockchain, artificial intelligence and Internet of Things.

[3] From 1980 to 2018, the purchasing power of the top 1% of global income increased by 80% to 240%, and that of the bottom 50% of global income increased by 60% to 120%; 44% of absolute income gained (AIG) has long been in the hands of the richest 5% of the world, while nearly 1/5 of total absolute income gain. In contrast, even the emerging middle-income classes of the globalization process together account for only 12–13% of global absolute income gain; in 2018, the top 10% as a proportion of national income was 34% in Europe, 41% in China, 46% in Russia, 48% in the US, 54% in Saharan Africa, 55% in India, 56% in Brazil and in the Middle East 64%. For systematic research, see Branko Milanovic, *Global Inequality: A New Approach for the Age of Globalization*, Cambridge, Massachusetts, and London: The Belknap Press of Harvard University, 2016; Branko Milanovic, *Capitalism, Aone: The Future of the System That Rules the World*, Cambridge, Massachusetts and London: The Belknap Press of Harvard University, 2016; Branko Milanovic, *Capitalism, Aone: The Future of the System That Rules the World, Cambridge*, Massachusetts and London: The Belknap Press of Harvard University, 2019; Thomas Piketty, *Capital and Ideology*, Cambridge, Massachusetts and London: The Belknap Press of Harvard University, 2020.

[4] Laura Tyson, Michael Spencer, 'The Impact of Technology on Income and Wealth Inequality', in [US] Heather Bussey, Bradford DeLong, Marshall Steinbaum, eds, After

in the United States over long periods points to the frequent disconnects between enormous technological advances and human adaptations to them, and the crises these disconnects provoke, which continue to drive dramatic changes in ethics, religious beliefs and political systems.[5] Carles Boix illustrates that the rapid development of information technology is exacerbating the inherent tension between democracy and the market, that digital technology no longer directly benefits the majority as technological progress did in previous historical periods, and that the political responsibility now lies in preventing elite self-imposed isolation, maintaining a functional market and compensating those who suffer.[6] There has also been considerable research by domestic scholars in recent years on the disparities in development capacity and the uneven distribution of wealth resulting from digital technology.[7] The warning issued by economic historian Carl Benedikt Frey can be seen as a problem-aware summary of the established research. If left unaddressed, the social costs of the widening divide between winners and losers generated by technological progress could far outweigh the individual burdens resulting from the direct impact on jobs; a worsening economic divide would translate into a larger political divide that will challenge the governance structures of the human world.[8]

However, most of the research has been conducted from the outside of technology. It has failed to address the 'selectivity of Technology Route': Can new Technology Routes be chosen to promote the participation of the majority in development more strongly? In the recent past, technology elites and business people have put forward many ethical ideas to respond to social anxieties, such as 'technology for good,' 'algorithmic transparency' and ethical norms for artificial intelligence. However, these

Piketty: A New Agenda for Inequality Research, translated by Yu Jiang and Gao Desheng, Peking: CITIC Press, 2022, p. 179.

[5] Robert William Fogel, *The Fourth Great Awakening and The Future of Egalitarianism*, Chicago: University Of Chicago Press, 2002.

[6] Carles Boix, *Democratic Capitalism at the Crossroads: Technological Change and the Future of Politics*, Princeton and Oxford: Princeton University Press, 2019.

[7] For example, Qiu Zeqi et al., From the Digital Divide to Dividend Differences - A Perspective on Internet Capital, *Chinese Social Sciences*, No. 10, 2016; Qiu Zeqi and Qiao Tianyu, Technological Changes in E-Commerce and Farmers' Co-Development', Chinese Social Sciences, No. 10, 2021.

[8] Carl Benedikt Frey, *The Technology Trap: Capital, Labor, and Power in the Age of Automation*, Princeton &Oxford: Princeton University Press, 2019. p. 343.

ideas do not touch the 'winner-takes-all effect.' We believe that it is more important to consider how to guarantee the autonomous participation of ordinary people, including most working population, in the development process, so that technology can truly become a tool for the inclusive development of human society. Only by making technological progress subservient to the development needs of the majority of ordinary people, including the working population, can we effectively achieve relatively balanced wealth growth.[9]

The Communist Party of China has proposed the development of digital technology and the digital economy as a national strategy at a time of dramatic changes in the development situation at home and abroad, attaching great importance to the role of the healthy development of the digital economy in promoting the construction of a new development pattern, the building of a modern economic system and the construction of a new national competitive advantage.[10] In China, the development and growth of the digital economy need to be emphasized under the strategic concept of accelerating the resolution of unbalanced and inadequate development and promoting common prosperity in phases. From the strategy of common prosperity, we must face the serious problems of the current mainstream Technology Route and explore institutional and technological innovations to create a 'development environment for all' in the digital age. The second part of the article theorizes the meaning of 'Technology Route selectivity' and discusses the state's responsibility in the selection process. The fourth part discusses the theoretical value and practical possibilities of the new Technology Route while revealing its policy implications in light of the emergence of new technologies in practice.

[9] Regarding its intellectual origins, this chapter attempts to continue Proudhon's understanding of the machine as both a product and an antithesis of the division of labor, which should and could make the machine a 'symbol of human freedom.' See Proudhon, *The Philosophy of Poverty*, vol. 1, translated by Yu, Shutong and Wang, Xuehua, Peking: The Commercial Press, 2010, pp. 164–177.

[10] Xi, Jinping, Keep making China's digital economy stronger, better and bigger, Seeking Truth, 2022, no. 2. In Seeking Truth: http://www.qstheory.cn/dukan/qs/2022-01/15/c_1128261632.htm.

Technology Route Optionality and the Responsibility of the State

(1) The theoretical connotation of the selectivity of the Technology Route. There are always different attitudes toward technological progress: optimism and pessimism. Technological optimism asserts that technological change is the driving force behind economic and social development and that new technologies will lead to productivity gains and the creation of new jobs[11] while in the view of technological pessimism, the replacement of labor by technology is the source of social unrest, and technological progress often does not lead to social development but instead masks deep-seated contradictions, with the recurrence of the destruction of the means of production in various places at different historical periods. The 'Luddist movement' (Luddism), which recurred in various parts of history to destroy the tools of production, is clear evidence of the resistance of workers to technological oppression.[12] However, there is a common blind spot in these two contrasting attitudes that have not been identified: whether optimistic or pessimistic, there is a risk of viewing technology from the outside, of failing to open the 'black box' of technological progress and thus falling into 'technological determinism.' This is a common blind spot: whether optimistic or pessimistic, technology may be viewed from the outside, failing to open the 'black box' of technological progress and thus falling into 'technological determinism.'

The diversity of Technology Route possibilities. Opening the 'black box' of technological progress reveals that more than one technological solution exists in a given environment and that different Technology Routes allow for different adjustments in the mix of input factors (such as labor and capital). Jon Elster, the representative scholar of Western analytical Marxism, has criticized scholars of the so-called production process theory for their mistaken belief that there is only one feasible Technology Route in a given environment, which determines the constant

[11] For an optimistic discourse on information technologies such as artificial intelligence in recent years, see the work of two MIT economists, Erik Brynjolfsson and Andrew McAfee, *The Second Machine Age: Work, Progress, and Prosperity in a Time of Brilliant Technologies*, New York: W. W. Norton &Company, Inc., 2014.

[12] See, for example, Nigel Cameron, Will Machines Take Your Job? translated by Wei Qian and Wang Litao, Peking: China Workers Press, 2020.

ratio of the factors of production.[13] According to these scholars, Marx also had a determinist position on technology. When he recognized that the machine was a tool of the capitalists to exploit the workers, the only response he could offer was the one-size-fits-all means of radically changing the relations of production: the working class taking possession of the means of production and socializing production.[14] However, Elster points out that there are multiple technical realization options that coexist and can establish relationships between many different factors of production. For example, in the production of nitrogen fertilizer through the electrolysis of water, producers can choose a combination of factors with more capital (electrolysis chambers) and less energy (electricity), or they can choose a combination of factors with less capital and more energy, and the interrelationship between labor and capital is different under the different options.[15] Without commenting on the validity of Elster's criticism of production process theory per se, I would like to highlight the diversity of Technology Route that it reveals, which broadens the horizon in which technological progress can be viewed and the space in which public policy can be implemented. In other words, it is possible to use policy instruments to change technological solutions and thus the matching ratios of inputs to factors of production, and accordingly to adjust the relative position of workers in the production process while maintaining the same output targets at a certain level of technology.[16]

[13] Jon Elster, *Explaining Technical Change: A Case Study in the Philosophy of Science*, Cambridge: Cambridge University Press, 1983, p. 163.

[14] For a critique of the tendency toward 'technological determinism' in modern political theory, see William Lazonick, *Competitive Advantage on the Shop Floor*, Cambridge: Harvard University Press, 1990.

[15] Jon Elster, *Explaining Technical Change: A Case Study in the philosophy of Science*, Cambridge: Cambridge University Press, 1983, p. 94.

[16] US economists have recognized that skill-biased and labor-saving technological change has replaced manual labor in routine blue- and white-collar work tasks and has contributed significantly to the stagnation of real wages in the general US labor force, the decline in real earnings limits in the non-college-educated workforce and the decline in union penetration and bargaining power, but note that 'empirical data suggests that the elasticity of substitution of capital and labor has historically been significantly less than one' (Laura Tyson, Michael Spencer, The Impact of Technology on Income and Wealth Inequality, in Heather Bussey, Bradford DeLong, Marshall Steinbaum, eds, *After Piketty: A New Agenda for Inequality Research*, translated by Yu, Jiang and Gao, Desheng, Peking: CITIC). This is a striking manifestation of what Richard G. Lipsey has criticized, and thus risks falling into 'technological determinism': economists' focus on

The process of technological development is a social process. Opening the 'black box' of technological progress requires that it can be seen in terms of the social factors that influence its direction rather than simply as a result of scientific and technological principles. As David F. Noble, a historian of technology at MIT, has analyzed: 'For the process of technological development is essentially social, and thus there is always a Beyond the very real constraints of energy and matter exists a realm in which human thoughts and actions remain decisive. It merely consists of an evolving range of possibilities from which people choose A social history of technology that explores beneath the appearance of necessity to illuminate these possibilities which technology embodies. reveals as well the contours of the society that realizes or denies them.'[17] Noble's research on the history of the development of automation technology points to the existence of two different automated Technology Routes in the United States around the 1970s, correspondingly predicting two different socio-political consequences. One was a 'record-playback' method of recording the movements of skilled mechanics, issuing instructions to machines based on the records and continually adjusting them as they were applied, and the other was a 'program-control' method in which engineers programmed and controlled the production process in advance. The former method relies more on skilled labor and is more integrated into the actual labor production process. In contrast, the latter method meets the needs of management and the scientific community to control technical deviations and regulate the production process, and is more detached from the actual labor production process. The result of the social choice of Technology Route for automation is a reflection of the contrasting social forces: as the military's demand for standardized, high-volume output dominates the overall market demand, the Technology Route, which is more detached from the labor production process and

static data prevents them from seeing the dynamic and complex impact of the process of technological progress on economic growth. Richard G. Lipsey, Kenneth I. Carlaw, and Clifford T. Bekar, *Economic Transformations: General Purpose Technologies and Long-Term Economic Growth*, New York: Oxford University Press. New York: Oxford University Press, 2005, p. 8. The discussion later in this chapter seeks to show that different technological solutions have the potential to achieve different interrelationships between capital and labor.

[17] David F. Noble, *Forces of Production: A Social History of Industrial Automation*, New Brunswick and London: Transaction Publishers, 2011, preface xiii.

thus more conducive to management control of errors and management of workers, becomes dominant.[18]

These two points together form the core of the 'Technology Route optionality.' This understanding rejects technological determinist thinking and reveals that underneath the manifestations of technological progress, technological change and technological domination lies a complex contrast of social forces and an unbalanced power distribution structure. Technological progress is not a natural consequence of the application of scientific and technological principles; human initiative can have an impact on the direction of technological development; and what is needed to develop technology that benefits the majority of ordinary people is the formation of checks and balances on the dominant technological and business elites. Accordingly, it can be understood that the question of technology is a political one: it is the political responsibility of the state to explore options for technological progress that will help the development of ordinary people, including the working population at large—technology is important. It cannot be left to technologists and business people alone. Under conditions of modernity, the state has the political responsibility to reorient technological development. It should aim to build a creative and inclusive model of economic and social development, shaping its political philosophy on technological inventions and the prospects for their application.

(2) National responsibility for Technology Route selection. Historically, many examples of countries have successfully steered the development of Technology Routes. In the eighteenth century, Britain invented new technologies to save labor and increase energy efficiency, which led to a rapid increase in industrial productivity. However, the British Technology Route was not suitable for continental European countries. France, for example, with its labor-intensive, decentralized rural industries and the extremely high cost of coal mining, needed to be suited

[18] David F. Noble, *Forces of Production: A Social History of Industrial Automation*, New Brunswick and London: Transaction Publishers, 2011, p. 84. This socially selective outcome was not necessarily the ideal situation, and in the early 1980s, there was strong criticism in American society that 'new jobs created by high technology were far fewer than those lost in manufacturing,' denouncing the negative effects of labor-replacing automation technologies. 'Anlerica Rushes to High Tech for Growth', *Business Week* (March 28, 1983, cited in David F. Noble, *Forces of Production: A Social History of Industrial Automation, New Brunswick, UK). History of Industrial Automation*, New Brunswick and London: Transaction Publishers, 2011, p. 348.

to the widespread adoption of the new British inventions. After briefly promoting new British technologies such as the Jenny spinning machine in the late eighteenth century, the French government turned to set up subsidies and other mechanisms to support national companies exploring new Technology Routes. The Jacquard spinning machine was suitable for decentralized production methods, but also allowed for a degree of standardized batch production through programmed control through perforated paper tapes that recorded the movements of skilled spinners.[19] After the introduction of the Jacquard spinning machine, the new invention was quickly adopted by most small and medium-sized cotton spinning businesses and cottage industries in rural and urban France, and the then Emperor Napoléon Bonaparte visited the Jacquard spinning process in 1805 and gave the new machine his great support. According to statistics, there were only about 1,000 jenny spinning machines in all of France (less than 5% of the British stock), which were used to produce large industries in the major cities. The mechanization of the cotton textile industry in France was mainly achieved by applying of national technical inventions.[20] Fernand Braudel, the historian of the Chronicle school, later noted that the unique technological path maintained a more inclusive economic and social development in France, with the simultaneous development of the large industry as an industrial superstructure and small industry, which had long been the majority in the countryside scattered throughout the country (what French economists at the time called 'double progress'), with the majority of the workforce in active employment, with most of the workforce working in active small-scale

[19] The Jacquard Loom was a mechanical technology that recorded the movements of skilled spinners on perforated paper tape, which was then used to control the operation of the machine, relying on skilled workers and enhancing the ability of new workers to operate it. The Jacquard spinning machine is considered a pioneer of the modern computer. See Ylva Fernaeus, Martin Jonsson, Jakob Tholander, Revisiting the Jacquard Loom: Threads of History and Current Patterns in HCI. CHI 2012, May 5–10, 2012, Austin, Texas, USA. Cited from Cui, Zhiyuan, Marx, Tocqueville and Proudhon in 1848, The twenty-first Century (Chinese University of Hong Kong), 2018 (6).

[20] For a discussion of the economic and social conditions that enabled the creation of the new technologies in Britain during the Industrial Revolution, as well as their application and transformation in various countries such as France, India and China, see Robert Allen, *The British Industrial Revolution in Global Perspective*, translated by Mao, Likun, Hangzhou: Zhejiang University Press, 2012. For information on technological inventions and applications in cotton textiles, see pp. 279–334 of the book.

urban and rural enterprises and with greater opportunities to set up their businesses.[21]

The first volume of Capital and the classic works such as the Condition of the Working Class in England has revealed that the early industrialization of Britain had a heavy social cost: firstly, the small farmers and small industrial production methods had no room to survive under the rapid development of large industries, and a large number of landless peasants and bankrupt small operators had to enter large factories as workers on the assembly line; secondly, the workers had no bargaining power in front of the capitalists who used large machines, and the industrial workers were regarded as 'subordinate' to the machines. Secondly, the workers had no bargaining power in front of the capitalists who used the big machines, and the industrial workers were regarded as 'subordinate' to the machines. The British path was not followed by the later world economic powers, such as the United States, which achieved a temporary settlement after the bitter conflict between professional unions and employers in the mid- to late nineteenth century, when the federal government offered support to entrepreneurs in the direction of technological development in exchange for concessions from the business community to high wages and benefits for skilled workers (the adoption of a pension system in the late 1930s was the most important achievement of the trade union movement). In the late twentieth century, a model of technological innovation based on a free labor market, a free technology transfer system and a venture capital system was gradually established, characterized by the disruption of the existing Technology Route; another example is Germany, which actively promoted the organizational change of the craft sector in the early years of industrialization and gave them exclusive rights to skills certification, forming a stable system of internal skills training, and in the second half of the twentieth century, a model of STI was developed that facilitated the improvement and consolidation of the existing Technology Route.[22]

21 Braudel, Fernand, *The Character of France: People and Things (vol. 2)*, translated by Gu, Liang and Zhang, Zeqian, Peking: The Commercial Press, 1997, pp. 249–265.

22 The history of institutional evolution and technological innovation in the major countries of the capitalist world has developed two important overlapping but distinctive literature. The classic work, known as 'Diversity in *Capitalism*,' focuses on the differences between national systems and is Peter A. Hall and David Soskice, *Varieties of Capitalism: The Institutional The classic work is Peter A. Hall and David Soskice, Varieties of Capitalism: The Institutional Foundations of Comparative Advantage*, Oxford: Oxford

In conclusion, historical experience shows that the core of independent development must be to create, at the right time, a technological development path that is appropriate to the country's political and economic system, market structure, workforce composition and cultural traditions by strengthening the country's capacity to choose its own Technology Route.

In today's world, many developing countries have adopted the US development paradigm and Technology Route as a model. However, is the US-led Technology Route universally valid and guaranteed to lead to sustainable economic growth and equitable wealth distribution? The answer, of course, is no, and the brief overview in Part III of this chapter will reveal the problems underlying the information technology revolution that emerged in the US. With the world now during dramatic change and China at a critical stage constructing a new development landscape, it is incumbent upon us to think calmly about the most appropriate technological path for us.

China has a sound industrial chain and an industrial ecology capable of supporting the production of most industrial products, clothing and footwear to aerospace and from raw material processing to industrial mother machinery manufacturing. The two tasks of eliminating high energy consumption, high pollution and low output industrial enterprises and improving the efficiency of innovation and leadership in high-end industries are both of considerable urgency. At the same time, China faces the dual policy needs of achieving sustained growth in labor income and expanding and strengthening advanced manufacturing to enhance industrial competitiveness. The Technology Route we have adopted is incompatible with our industrial ecology and workforce structure if we pursue a transformation to frontier innovation industries and advocate automation as the core orientation in industrial enterprises (forming technological innovations of machines for humans, less human and unmanned). In addition, although in recent years, the national level has proposed the 'Great Artisans' strategy for a strong manufacturing industry, the implementation of the policy is mainly aimed at the demand

University Press, 2001. Harry Braverman, *Labour and Monopoly Capital: The Degradation of Work in the Twentieth Century*, New York and London: Monthly Review Press, 1974; Michael Burawoy, The Politics of Labor. Burawoy, *The Politics of Production: Factory Regimes Under Capitalism and Socialism*, London: Verso, 1985; Kathleen Thelen, *How Institutions Evolve: The Political Economy of Skills in Germany, Britain, the United States, and Japan*, New York: Cambridge University Press, 2004.

of large enterprises for special skills and high-quality jobs. At the same time, the majority of the population, the general workforce and the rest of society, should be supported with skills training and resources for their development aspirations. Various sectors still need to pay more attention.[23] These are complex realities that need to be properly considered. In any case, there is a need for a fundamental study to reflect on the information technology revolution that emerged in the United States at a time when the development of digital technology and the growth of the digital economy are national strategies and to establish a basic consensus on a clear national choice of Technology Route that will facilitate the development of everyone in the digital age.

The 'Elitism' of the IT Revolution Technology Route

To reflect on the current development of information technology from the perspective of the selectivity of the Technology Route is not to accuse technology of conspiring with capital and power to erode the public. The question of how to make technological progress subservient to the development needs of ordinary people, including the majority of working people (the 'popular route' of technological development), is worth discussing. From this perspective, the information technology revolution that emerged in the United States has followed an 'elitist' Technology Route, which has prevented its innovative ideas and creative potential from spreading to a broader range of industries and types of businesses, and has made it challenging to expand employment and entrepreneurial opportunities for the working population and other ordinary people.

(1) Starting with the 'difficulty of digital transformation.' In current public opinion, the 'difficulty in digital transformation' of SMEs in traditional manufacturing and other industries is generally understood as an operational problem of promoting the application and diffusion of digital technologies in enterprises. For example, a report hosted by a media outlet states[24]: 'According to data from the fourth national

[23] Feng, Kaidong and Li, Junran, The political economy of skills: three sets of key propositions, *BYU Political Science Review* (4th series), Vol. 2, 2018, pp. 198–200.

[24] Industrial Internet Talk: The Key to Cracking the Difficulty of Digital Transformation of SMEs is to Make Up for the Shortage of Talents, https://tech.china.com/article/20210918/092021_879389.html.

economic census, 89% of China's more than 2,000 small and medium-sized enterprises are in the exploration stage of digital transformation, only in the core aspects of design, production, logistics, sales and service; 8% of enterprises are in the practice stage of digital transformation, digitally transforming core equipment and business data for digital transformation; and only 3% of SMEs are in the deep application stage of digital transformation.' The report analyzes the reasons for the difficulty of digital transformation of SMEs; one is the weakness of data collection capability, low level of technology application and other basic capabilities, and the second is the high cost of transformation and limited resource investment, the most important of which is 'insufficient supply of talents': data from the China Industrial Internet Research Institute shows that more than 75% of SMEs have less than 20% of digital transformation talents.

However, is digital transformation difficult only because of problems on the part of real businesses? Why do current digital technologies fail to meet the transformation and development needs of different types of businesses? Can digital technologies be developed based on traditional manufacturing needs and various types of MSMEs? These deeper questions should be addressed. A market-based consultancy recently found that 'MSMEs lack a stronger endogenous demand for digital transformation than large enterprises, and their revenue levels are not sufficient to support systematic digital transformation. Instead, MSMEs are more market-dependent, and their digital upgrading model should first address their external ecosystem and then be driven from the outside in.'[25] While still not noting the different routes of digital technology, this observation starts from within the enterprise and raises the issue of the different needs of different enterprises.

At present, although the high penetration of digital products gives the illusion that the IT revolution has reached all areas of life, the use of digital products is not the same as being able to participate in the development of the digital economy. The inability of companies in the real economy, such as the manufacturing sector, to use digital technology in depth (e.g., computers are only used in areas such as office management processes) is due in large part to the 'elitist' orientation of the current

[25] Epidemic Watch: Digital Upgrade Research Report for Chinese SMEs (2020), http://www.cbdio.com/BigData/2020-04/13/content_6155587.htm.

mainstream development of digital technology, which makes it difficult to take into account the specific needs of these companies.

A pioneer of the new technology wave, Peter Thiel, founder of PayPal and Facebook's first outside investor, famously criticized that we wanted a car that flies, and all we get is 140 characters.[26] The criticism of '140 characters,' which refers to the maximum number of characters per Facebook post, points to the gradual fragmentation of the ICT industry that emerged in the United States from the needs of its economic and social development. The reasons for this fragmentation are complex.[27] We will only sketch this situation in terms of the basic state of work of the information industry sector: microcomputers appeared in the late 1970s, independent software suppliers became the largest market supplier, and after about 1982, the invention and popularity of the PC stimulated even more the explosive growth of the software industry—'software defines everything.' 'Software swallows the world' became the most convenient way of working and ideal for the ICT sector, and software developers mostly came from the original computer centers, or were students straight out of university, unaware of the experience of the working population in other industries and the development needs of other businesses, and do not need to go deep into production processes and work sites in order to do their job.[28] Karls-Boikos sees this fragmentation as a detriment to a

[26] William B. Bonvillian and Peter L. Singer, *Advanced Manufacturing: The New American Innovation Policies*, Cambridge: The MIT Press, 2017, p. 188.

[27] For an analysis of the reasons for the 'elitist' orientation of the IT revolution at different levels, see William Lazonick, *Sustainable Prosperity in the New Economy? and High-Tech Employment in the United States*, Michigan: W.E. Upjohn Institute for Employment Research, 2009 (discussing the ways in which US technology firms, in maximizing shareholder interest, have undermined the incentives for innovation that contribute to sustainable development), key incentives for innovation that contribute to sustainable development, such as financial commitment, organizational integration and insider control, as a significant cause of worsening income disparities; Paolo Bory, *The Internet Myth: From the Internet Imaginary to Network Ideologies*, London: University of Westminster Press, 2020 (discussing how the discourse of the technological elite has gradually become the dominant narrative shaping the direction of the Internet during its development). For a critique of the current European and American legal systems' protection of the 'elitist' route to information technology, see Amy Kapczynski, The Law of Informational Capitalism, *The Yale Law* Journal, Vol. 129, No. 5 (2020). 5 (2020).

[28] Alfred D. Chandler, James W. Cortada, eds, *Information Changed America: The Forces Driving National Transformation*, translated by Wan, Yan and Qiu, Yanjuan, Shanghai: Shanghai Far Eastern Publishing House, 2011, pp. 290–291; Margaret O'Mara,

'democratic' form of life: innovation in a company under traditional technology requires experienced mechanics and engineers with appropriate knowledge of existing production processes to try things out in the workshop, thus respecting the opinions of frontline workers to a certain extent in the process of technological innovation, whereas under IT conditions, a group of leading physicists, mathematicians and software engineers, for example, can set the direction of technological innovation in a research laboratory far from the workshop.[29]

This detachment from productive labor and business practices led to a great deal of skepticism when information technology, such as computers and the Internet, became widespread in the United States in the late 1960s: the innovations it generated have been concentrated in the areas of entertainment, information and communication, but have failed to change the established ways of production and life in many more areas.[30] As discussed later in this chapter, the 'elitist' orientation of the Technology Route in the ICT industry is most notable for its lack of direct and immediate contribution to the traditional real economy, such as manufacturing. Despite the continued decline, traditional sectors of the economy, such as manufacturing, still accounted for more than two-thirds of the US GDP in 2003. By 2019, the value added of the US manufacturing sector amounted to US$2.34 trillion, or 10.9 percent of GDP, while the ICT industry's share of US GDP rose by only about 1 percent from 2000 to 2018, with a value of about US$1.72 trillion, or 8 percent of GDP, in 2018, trillion USD and 8.3% of GDP.[31] However, under the overwhelming discourse of the rapid development of new economic industries, such as information technology, the development and innovation needs of traditional industries, such as manufacturing, have not received the attention they deserve. The current study on the 'difficulty of

The Silicon Valley Code: How Technological Innovation Is Reshaping America, translated by Xie, Nijie, Peking: CITIC Publishing House, 2022, pp. 407–420.

29 Carles Boix, *Democratic Capitalism at the Crossroads: Technological Change and the Future of Politics*, Princeton and Oxford: Princeton University Press, 2019, pp. 177–215.

30 Robert ˙Gordon, *The Rise and Fall of American Growth*, translated by Zhang, Linshan et al., Peking: CITIC Publishing Group, 2018, pp. 552–560.

31 See Han, Wenyan, Xiong, Yonglan and Zhang, Zhiqiang, Analysis of the Development Trend of the US Information and Communication Industry in the Last 20 Years and Implications, World Science and Technology Research and Development, No. 2, 2021; Guoxin Securities, 'Overseas Market Week Overview: Mining the US Manufacturing Data', 5 January 2021, https://pdf.dfcfw.com/pdf/H3_AP202101061447750845_1.pdf.

digital transformation' of small and medium-sized enterprises in the real sector in China is in danger of falling into the 'paradox of the information technology revolution' in the United States. It requires a high degree of vigilance: digital technology, which has failed to lead deep economic and social change, has occupied the core of policy and public opinion discourse. Digital technology, which has failed to lead deep economic and social change, has taken center stage in policy and public opinion discourse and more resources and institutional guarantees have, in turn, made the IT revolution more elitist and self-referential.[32]

(2) The 'elitist' approach to information technology and the decline of American manufacturing. The US federal government's aim in promoting the IT revolution was to produce a new generation of products rather than to enhance the competitiveness of the manufacturing industry, so from the beginning it did not take into account the needs of the working population in the actual manufacturing industry for practical experience.[33] Out of the need to compete with the Soviet Union in science and technology after World War II, the US spawned an information technology revolution led by the Department of Defence agencies. For example, after the Soviet Union launched the first human satellite in 1957, the US established the Defense Advanced Research Projects Agency to fund scientists at various universities and research institutions to freely engage in scientific research and translate the results, and also invested the Computer Science Division to provide early research support for start-ups and promote semiconductor research and human–computer interaction, leading to the early development of the Internet.[34] While vigorously promoting the flourishing of the information industry, due to an overly optimistic belief that the benefits of technological innovation could compensate for the stagnation of physical manufacturing industries or their relocation overseas, US industrial policy from the mid-1970s was

[32] Francesco D. Sandulli, Solving the Paradoxes of the Information Technology Revolution: Productivity and Inequality, https://papers.ssrn.com/sol3/papers.cfm?abstract_id=1,192,568.

[33] William B. Bonvillian and Peter L. Singer, *Advanced Manufacturing: The New American Innovation Policies*, Cambridge: The MIT Press, 2017, p. 34.

[34] Mariana Mazzucato, *Innovative government: Building a Symbiotic and Win–Win Relationship Between the Public and Private Sectors*, translated by Lei Li, Dongxin Shuang and Shanjian Cheng, Peking: CITIC Publishing House, 2019, pp. 90–101. On the history of the development of the Internet, see Janet Abbate, *Inventing the Internet*, Cambridge, Mass.: MIT Press, 1999.

biased toward supporting high technology and information technology industries that could not expand employment on a large scale, as two MIT industrial policy experts noted: 'the United States is not supporting innovation in significant job-creating sectors.'[35]

In 1987, Robert M. Solow, the Nobel Prize in Economics winner that year, famously questioned the then burgeoning IT industry with his 'productivity paradox': 'Computers are everywhere, except in productivity statistics.'[36] Solow's observation is supported by changes in productivity data in the US in the following decades, namely that the rapid growth in global top-level IT R&D was accompanied by a significant decline in productivity growth in the US manufacturing sector as a whole: from 2004 to 2014, total factor productivity grew at an average annual rate of 0.4%, the lowest in the country's history since 1890; from 2005 to 2016, total US labor productivity grew at an average annual rate of 1.3%, less than half of the 2.8% average annual growth rate of the previous decade (1995 to 2004). A more direct indication of how little IT has helped at least the physical manufacturing sector to increase productivity is a figure from the US Bureau of Labor Statistics: US manufacturing productivity grew at an average annual rate of 4.1% between 1989 and 2000, while between 2007 and 2014, manufacturing productivity, which absorbed the benefits of the IT revolution, fell to an average annual growth rate of 1.7%.[37]

The decline in productivity growth is not only the result of technological development (the investment of resources into economic activities that do not generate value output). However, it is also a consequence of using technology. Economist Robert J. Gordon explains the above change in productivity growth. While acknowledging that the mass introduction of information technology into production and life had initially increased efficiency in all areas, Gordon goes on to point out that the scope and depth of the current information technology revolution are minimal compared to the second Industrial Revolution, which changed

[35] William B. Bonvillian and Peter L. Singer, *Advanced Manufacturing: The New American Innovation Policies*, Cambridge: The MIT Press, 2017, p. 215.

[36] Robert Solow, We'd better watch out, *New York Times Book Review*, July 12, 1987.

[37] Robert Gordon, *The Rise and Fall of American Growth*, translated by Zhang, Linshan et al., Peking: CITIC Publishing Group, 2018, pp. 550–551; William B. Bonvillian and Peter L. Singer, *Advanced Manufacturing: The New American Innovation Policies*, Cambridge: The MIT Press, 2017, p. 54.

all significant aspects of production and lifestyle around the world. For example, smart phones have so far been used mainly for personal activities in the office; e-commerce is too low a proportion of the overall economy; and the integration of digital technology into the financial and banking sectors has not reduced the reliance of businesses on professionals such as financial accountants or offset the willingness of banks to expand their physical branches. In the words of Professor Gordon, 'Each of the sectors discussed above (office, retail, finance, banking) changed fundamentally and revolutionarily in the 1980s and 1990s. It is only in the last decade that computer hardware, software and business methods have ossified into a set of slowly changing conventions.'[38]

One computer researcher blamed federal policies for the lack of adaptation to workers and their skills when the US Internet economic bubble collapsed in 2000: 'The huge investments made in infrastructure over the last few decades have taught us how to move money, physical entities, information and a growing number of jobs quickly, but we have been unable to deploy workers and their skills at the same pace.'[39] Industrial policy experts are more pointed in pointing out that different industries and their various types of businesses have their own relatively unique and well-functioning technological solutions, business models and human capital structures and that it is highly unreasonable and unlikely that a radical change in these well-functioning uniqueness to accommodate current information technology will be required.[40] Taken together, these criticisms lead to a lesson that needs to be reflected upon: the IT revolution, taught by the elitist Technology Route, is divorced from the practices of actual industrial production and thus offers little positive help in the face of the irreversible decline of the US domestic manufacturing industry into the twenty-first century.[41]

[38] Robert Gordon, *The Rise and Fall of American Growth*, translated by Zhang, Linshan et al., Peking: CITIC Publishing Group, 2018, p. 558.

[39] Bruce Abramson, *The Digital Phoenix: Why the Information Economy Can Be Reborn*, translated by Zhao, Pei and Zheng, Xiaoping, Shanghai: Shanghai Far Eastern Publishing House, 2008, p. 253.

[40] William B. Bonvillian and Peter L. Singer, *Advanced Manufacturing: The New American Innovation Policies*, Cambridge: The MIT Press, 2017, pp. 4–5.

[41] For an analysis of data on the reduction of US manufacturing, see Susan N. Houseman, Understanding the Decline of US Manufacturing Employment, Upjohn Institute Working Paper 18–287, https://research.upjohn.org/cgi/viewcontent.cgi?article=1305&context=up_workingpapers. Examples of the rapid decline of the US

(3) Problems with the 'elitist' Technology Route. It should be pointed out immediately that criticism of the current 'elitist' Technology Route in the ICT industry does not deny the innovative ideas and potential of the new technological revolution. In other words, the 'elitist' Technology Route is not incapable of generating innovative ideas and creativity but is perhaps more accurately characterized by the terminology of Professor Roberto M. Unger of Harvard Law School, who describes the current rapid development of digital, intelligent and new technologies as a 'new technology.'[42] On the one hand, digital technologies offer the promise of low-cost sharing of resources and information, the pooling of mass wisdom and the disruption of the conventional division of labor, offering the prospect of conceptual vanguardism, but on the other hand, because of their Technology Route. On the other hand, the elitist orientation of the Technology Route, which is divorced from the majority of productive labor experience and the specific needs of firms, prevents the widespread diffusion of innovation in practice.

As a result, the 'elitist' approach of the IT revolution has limited the innovation and creativity contained in the new technologies to a few frontier industries and a few frontier companies, creating a 'winner-takes-all effect' in economic and social development. For example: firstly, the expansion effect is created by digital technology. In other words, development based on digital technology does not overcome diminishing marginal returns through 'creative destruction' (innovation), as Joseph A. Schumpeter put it, but through the continuous acceleration of factor inputs.[43] The second is the creation of an exclusionary effect through digital technology. This is the inevitable result of rising returns to scale,

telecommunications equipment industry in the face of rapid ICT development include the bankruptcy of the former telecommunications equipment giant Nortel in 2008 and the failure of the U.S. telecommunications equipment industry. Nortel's bankruptcy in 2008 and Lucent's sale to France's Alcatel (later acquired by Finland's Nokia) shortly afterward also illustrate the damage that federal industrial policies, regulations and corporate strategies around the elitist Technology Route have done to the telecoms equipment industry. Atkinson, Who Lost Lucent? The Decline of America's Telecom Equipment Industry, https://americanaffairsjournal.org/2020/08/who-lost-lucent-the-decline-of-americas-telecom-equipment-industry/.

[42] Roberto M. Unger, *The Knowledge Economy*, London and New York: Verso, 2019, p. 1.

[43] Roberto M. Unger, *The Knowledge Economy*, London and New York: Verso, 2019, p. 29.

making it inevitable that a small number of technology firms will become industry giants and that the formation rate of giant scale will become increasingly rapid. The resource extraction capacity and scientific research creation capacity of the giant platform determine its overwhelming advantage over other market players and exclusivity benefits.[44] Third, the formation of the centralization effect with the help of digital technology. This is mainly reflected in the increasing ability to make precise decisions from the top down and to create market demand as resource extraction and R&D capabilities increase, which allows giant companies to increasingly detach themselves from the real needs of the economy and society and to reconstruct a new ecology of all aspects of social life, acquiring the privilege of 'unilaterally claiming human experience as free raw material to be transformed into behavioral data.'[45] In a report published in 2015, the McKinsey Global Institute cites empirical research confirming this situation of digitization leading to a winner-takes-all effect: 'Digitization has made the competition more intense and rapid, and has increasingly created a winner-takes-all business environment. Since 2000, the average variance of North American companies' return on capital has been more than 60 percent higher than prevailing between 1965 and 1980. Not only are profits growing, but in some of the most digitally advanced industries, the profits of leading companies are greater than ever before. One of the biggest competitive shifts is the rise of the dominant digital platforms, whether e-commerce marketplaces, operating systems or social media networks.'[46]

[44] Large platforms such as Amazon can leverage their information management and resource allocation capabilities to implement a 'just-in-time' business model, dramatically reducing costs and making it completely impossible for small and medium-sized platforms to compete. Stanley, Madeleine Gabriel and Geoff Mulgan, Imagination Unleashed: Democratising the Knowledge Economy, Published by Nesta, March 2019. https://www.nesta.org.uk/report/imagination-unleashed/.

[45] Shoshana Zuboff, *The Age of Surveillance Capitalism: The Fight for a Human Future at the New Frontier of Power*, New York: Public Affairs, 2019, p. 8. In her book, Zuboff focuses on the critique of giant corporations such as Google that have developed 'surveillance capitalism'. Other literature provides richer theoretical clues to the ability of current big data technologies to shape markets, such as the differentiation of people into 'high-value' and 'high-risk' categories and discriminatory implementation strategies. The Law of Informational Capitalism, *The Yale Law* Journal, Vol. 129, No. 5 (2020), p. 1478.

[46] McKinsey Global Institute, Digital America: A tale of the haves and have-mores, December 1, 2015, https://www.mckinsey.com/industries/technology-media-and-telecommunications/our-insights/digital-america-a-tale-of-the-haves-and-have-mores.

Worldwide, new technologies have also had the effect of inducing a worsening of global inequality, as Dani Rodirik, professor of international economics at Harvard University, has pointed out that while the effects of new technologies driving industrial development and exports are increasingly evident in developing countries and regions involved in the formation of global value chains, global value chains have neither expanded employment nor contributed to the export-related. Instead, the traditional comparative advantage of these countries and regions in terms of unskilled labor has been eroded by the labor-saving and resource-saving effects of new technologies.[47] This lack of inclusive development has its roots in political economy: the current model of globalized development is the result of the internationalization of the Fordist mode of production.[48] Global value chains are based on an intra-product division of labor, with different labor tasks such as research and development, processing of raw materials, manufacturing of components, assembly of products and distribution being assigned to different countries and regions, sharing different shares of the value added of the product; the IT revolution, due to its elitist orientation away from the manufacturing production process, has not changed but rather strengthened this increasingly rigid division of labor, making it difficult for developing countries and regions to gain opportunities to upgrade their position in global value chains and increase their innovation capacity through their participation in the globalization process.[49]

[47] Dani Rodirik, New Technologies, Global Value Chains, and the Developing Economies, NBER WORKING PAPER SERIES, October 2018, http://www.nber.org/papers/w25164.

[48] Cui Zhiyuan, Double-cycling, Global Value Chains and Additive Manufacturing: Preliminary Theoretical Reflections on New Development Strategies, *21st Century* (Institute of Chinese Culture, Chinese University of Hong Kong), August 2021 issue, General 186, p. 40. For a discussion of the Fordist mode of production and the post-Fordist mode of production that challenged it, see Cui Zhiyuan, The An-gang Constitution and Post-Fordism, *Reading*, No. 3, 1996; James P. Womack, Daniel T. Jones and Daniel Roos, *The Machine That Changed the World*, New York: Free Press, 1990.

[49] China is one of the few exceptions to this rule: according to the World Bank's Global Value Chain Report: 2019, China's ICT exports have gradually moved up the global value chain since smartphones entered the global market in 2007. China's ICT exports have gradually upgraded their position in the global value chain since smartphones entered the global market in 2007. The reason for China's climb up the global value chain is related to a number of complex factors, such as China's vast consumer market for mobile phones, various supportive measures by local governments and the ecosystem changes

In summary, under the 'elitist' Technology Route, the knowledge accumulated in the production and life practices of workers and the general public (including the problems and experiences accumulated in the business and production processes of SMEs) on a global scale are not allowed to be refined, facilitated and developed. In contrast, digital technologies, which are difficult to integrate with a broader range of production and life scenarios, tend to accelerate the unequal distribution of wealth and undermine the enthusiasm of workers and SMEs for employment and entrepreneurship. The creation and application of digital technologies, which are difficult to integrate with a broader range of production and life scenarios, tend to accelerate the unequal distribution of wealth and undermine the enthusiasm of workers and SMEs for employment and entrepreneurship, leading to severe development imbalances and polarizing social divisions. China has achieved a miracle of economic growth through its great experiment of reform and opening up, under the existing globalized development model and Technology Route.[50] However, in the wake of the radical changes in the global development environment and the sudden outbreak of the new epidemic, development conditions have changed, and China's development strategy has shifted from an export-oriented to a 'double-cycle' strategy, with domestic demand as the starting and ending point. In this new situation, we urgently need to innovate the path of digital technology development and change the 'elitist' Technology Route to one that is more conducive to the growth and diffusion of new technologies across a wide range of economic sectors, empowering ordinary people, including workers, and enterprises of all kinds in different industries.

driven by the open source of the Android operating system and the semi-open source of the Apple operating system. Cui, Zhiyuan, Dual-Cycle, Global Value Chains and Additive Manufacturing: Preliminary Theoretical Reflections on New Development Strategies, *21st Century* (Institute of Chinese Culture, Chinese University of Hong Kong), August 2021 issue, Vol. 186, pp. 40–41.

[50] For an analysis of the complexities of China's participation in the formation of global value chains, in terms of wealth growth and the constraints imposed by the established global division of labor, see also Xing, Yuqing, *The Enigma of Chinese Exports: Decoding the 'Global Value Chain'*, Peking: Sanlian Bookstore, 2022.

The 'Popular Route' to Digital Technology Development

As the new generation of ICTs is still in a phase of rapid development and has not yet been fully defined by the complex influence of various social forces, our discussion needs to start with theoretical analysis. As the new generation of ICTs is still at a rapid stage of development and has not yet been fully defined by the complex influences of various social forces, we need to start our discussion with theoretical analysis and, in this way, explore the theoretical and practical implications of the new technologies that are emerging in reality.

(1) A conceptual understanding of the new Technology Route. Let's start with a clear conceptual understanding. Take artificial intelligence (AI) as an example. The technological development of AI has progressed in recent years from explicit programming to machine learning methods (a statistical technique that allows computers and algorithms to learn, predict and perform tasks from large amounts of data) and 'deep learning' methods (algorithms that use multi-layer processes such as neural networks to improve machine learning, statistical reasoning and optimized algorithms); the ability of AI to replace humans in repetitive, computable tasks has reached unprecedented levels. To explore the possibilities of a new type of AI Technology Route, it is necessary to respect the proper relationship between technological tools such as AI and people. The elitist Technology Route, with its 'software defines everything' approach and ideals, essentially distorts the relationship between technology and people: the idea that a set of external, abstract standards can guide all aspects of human production and life.

Based on natural philosophical research, Professor Roberto Unger suggests that machines can help humans with repetitive, calculable tasks, where they can increase the precision and intensity of their work, but they cannot fundamentally replace them. This is because the 'imagination' of the human mind is a function that no machine can replace:

> The hallmark of the imagination is its negative capability: the ability of the mind to distance itself from a phenomenon or state of affairs and then to subsume it under a range of transformative variations; to cast aside its settled methods and defy its present presuppositions the better to see to cast aside its settled methods and defy its present presuppositions the better to see something that it could not see before, and then retrospectively to

> develop the methods and formulate the presuppositions that make sense of an insight It is about vision. Machines cannot even in principle have this transgressive and visionary power. It is a power rooted in the most fundamental attribute of our humanity: our transcendence over all the finite It is a power rooted in the most fundamental attribute of our humanity: our transcendence over all the finite determinations of our existence, our inability to be contained within the conceptual and social worlds that we build and inhabit.[51]

Neglecting the function of the 'imaginative' dimension of the mind is a denial of human initiative and a suppression of the prospects for the full development of the human person. For example, in the age of mechanized industry, although classical development economics had recognized the role of 'education' as the basis for economic growth, but workers, who were seen as 'subordinate' to the machine, were not expected any 'education,'—possessing a 'disposition to obey, basic literacy and numeracy, and manual dexterity, especially hand–eye coordination' would suffice.[52] From this understanding of Unger's, the correct technological solution can only be that of a machine as a tool to help man enhance his ability to expand and realize his 'imagination.' It should be noted that the human 'imagination' itself is difficult to define precisely and that its boundaries and potential cannot be determined in advance, so any technological tool developed to assist in the expansion of the 'imagination' must be subordinated to the developmental needs of the human being in a specific context and be based on solving problems in a specific developmental environment. Any development of technological tools to support the expansion

[51] Roberto M. Unger, *The Knowledge Economy*, London and New York: Verso, 2019, pp. 43–44. The intellectual affinity between Unger and Proudhon is apparent. It can be contrasted with the discussion in Kan Proudhon, Proudhon, *The Philosophy of Poverty*, vol. 1, translated by Yu, Shutong and Wang, Xuehua, Peking: The Commercial Press, 2010, p. 196.

[52] Roberto M. Unger, *The Knowledge Economy*, London and New York: Verso, 2019, p. 40. Is 'education' merely a tool to improve the structure of human capital, or can it be a device for the development of the whole person? Consider Unger's view of education as follows: 'We can establish universally an education that recognizes in every child a tongue-tied prophet, and in the school the voice of the future, and that equips the mind to think beyond and against the established context of thought and of life as well as to move within it.' Roberto M. Unger, *The Religion of the Future*, Cambridge: Harvard University Press, 2014, p. 29.

of the 'imagination' must therefore be based on the developmental needs of people in specific contexts and on solving problems in specific developmental environments rather than being divorced from the productive reality of most ordinary people. This is the basic requirement that technological progress must be 'for the masses': any technological innovation or invention is divorced from the concrete production and living practices of ordinary people will not be able to provide practical assistance for the comprehensive development of human beings.

Noted economist Daron Acemoglu has focused on the relationship between AI and labor in recent years.[53] Based on empirical research, his reflections provide a concrete example of how we can better understand the above conceptual understanding. Asimoglu offers a critique of the mainstream AI Technology Route: the current mainstream AI Technology Route is oriented toward automation, tending to allow cheaper capital relative to labor (coalescing into machines) to replace a range of tasks previously performed by human beings, but automated AI external to labor production practices. However, automated artificial intelligence external to the practice of labor production has not significantly increased productivity to increase the demand for labor, with potentially disastrous consequences for income inequality and social cohesion.[54] Based on a Technology Route selectivity perspective, Asimoglu points to the possibility of the 'right type of AI': 'Since AI is not just a narrow set of technologies with specific, pre-determined applications and functionalities but a technology platform, it can be deployed for much more than automation; it can be used to restructure the production process in a way that creates many new, high-productivity tasks for labor. If this type of 'reinstating AI' is a possibility, there would be potentially large societal gains in terms of both improved productivity and greater labor demand (which will not only create more inclusive growth but also avoid the social problems created by joblessness and wage declines).'[55]

[53] Daron Acemoglu & Pascual Restrepo, Robots and Jobs: Evidence from US Labor Markets, https://conference.nber.org/confer/2017/PRs17/Acemoglu_Restrepo.pdf.

[54] Daron Acemoglu & Pascual Restrepo, The Wrong Kind of AI? Artificial Intelligence and The Future of Labor Demand, NBER WORKING PAPER SERIES, http://www.nber.org/papers/w25682.

[55] Daron Acemoglu & Pascual Restrepo, The Wrong Kind of AI? Artificial Intelligence and The Future of Labor Demand, NBER Working paper series, http://www.nber.org/papers/w25682.

Asimoglu envisages several new AI technology solutions. For example, in education, AI is used to collect data on the specific responses of students when they encounter difficulties in learning, providing information to assist in the implementation of personalized education; in health care, AI is used to collect and analyze information on medical patients to enhance the work of healthcare workers and other technicians; and in augmented reality, through the use of human–machine collaboration technologies, by using interactive interfaces, AI is used to increase the ability of humans to sense, monitor and control objects to help workers perform high-precision production and integrated design tasks. These programs align with the non-elitist Technology Route: AI does not aim to replace teachers, medical staff and workers, but rather to help the latter collect and analyze information more accurately and assist them in their ability to perform their professional work more creatively. Asimoglu provides valuable inspiration for advancing the new Technology Route for digital technology from conceptual awareness to practical implementation.[56]

(2) A paradigm of change in the route of digital technology implementation. Is it possible to change an established digital technology implementation route? Here is an example of a somewhat successful case to show that it is possible to change a digital technology implementation route based on the frontline production needs of a real manufacturing industry. Before presenting our case, it is essential to reiterate the meaning of 'Technology Route selectivity': there can be different technological implementations of the same technological principle, and different Technology Routes can have different economic and socio-political effects.

Case: Paper, food, building materials and other traditional process industries in the development process generally encountered severe

[56] Drawing on Asimoglu's thinking, we might devise solutions that change the implementation route of new digital technologies. For example, ChatGPT, the natural language processing tool powered by artificial intelligence technology, is generating much buzz. ChatGPT enables intelligent question and answer functionality and can therefore replace jobs such as human customer service. ChatGPT could also be adapted to categorize conversations by accurately capturing keywords enabling more profound analysis of large volumes of customer service questions and answers, thus facilitating direct attention to customer feedback and thus improving product quality on e-commerce platforms. The latter technology route is worth supporting.

constraints, such as (1) the production process relies on manual experience to control the operation, and the frontline skilled workers' ('teacher') experience is difficult to effectively refine and systematize, with the aging of skilled workers (young people are increasingly reluctant to enter the factory); (2) the essential characteristics of the industry are that the raw materials used for production are natural resources, but there is a need to produce standardized, uniformly performing products. As a result, it is difficult to ensure the effective composition of raw materials based on manual experience, which often results in excessive waste of raw materials and energy consumption; (3) once a company has achieved specific scale, the complexity of multiple production lines and the collaboration of multiple production factors are beyond the capacity of human beings to manage, and relying on skilled workers to operate at this time can lead to low productivity. Here, we see that the ideas promised by the rise of a new generation of ICTs, such as the low-cost sharing of resources and information, the pooling of mass intelligence and the disruption of conventional labor division models, can address these developmental pain points of traditional process industries. However, existing digital technologies do not offer such effective solutions, as the non-standardized, non-computable production processes, characteristic of the traditional process industries, do not meet the requirements of software programming work under the elitist Technology Route. In a word, software engineers away from the production floor do not understand the process, and existing digital technologies cannot be integrated into the production floor.

Based on its understanding of traditional processes, a technology company formed by paper professionals has worked to achieve a breakthrough in technical barriers by splitting industrial Internet and artificial intelligence technologies and embedding them in critical aspects of the production processes of these traditional process industries, providing software and hardware products and data services for the digital intelligence transformation of process-based manufacturing industries. For example, for a paper manufacturing company, the technology company embedded artificial intelligence technology in several production links to achieve comprehensive quality management of human, machine, material, method, environment and measurement; by repeatedly grinding with skilled workers and setting process parameters based on their experience, it established an industrial extensive data analysis platform to stabilize the process, while helping to analyze and quickly locate and dynamically adjust the effective composition of raw materials through real-time prediction and early warning of quality. Based on process characteristics, a series of data analysis methods and tools are set up at key points to identify energy saving improvement points and achieve overall energy use optimization of the pulp and paper machine. We see that in this series of solutions, digital technologies such as artificial intelligence and big data do not replace skilled workers but rather transforming them based on their experience, with the ultimate effect of improving workforce.

We do not intend to portray the above cases as examples of success with universal demonstrative effect. However, in any case, established digital technologies do not help traditional manufacturing as quickly as their advocates claim. It is the consensus of those who know manufacturing that existing digital technologies 'optimize processes, but do not transform them.'[57] Recognizing the optional nature of the Technology Route means that digital technology is no longer seen as a stable, universal technological tool and that the elite such as software engineers, no longer have the privilege of directing change away from the production floor, but are required to break down these digital technologies in terms of manufacturing processes and the experience of frontline skilled workers, and make them work for the actual production process itself. This may lead to the creation of new forms of digital technology operations and products, which express the so-called new technology for the masses.

(3) The emergence of a new technology route and the direction of public policy. In reality, new digital technologies have emerged that are integrated with manufacturing processes and are more likely to guarantee participatory development in the digital age, enhancing the development capacities and opportunities of the working population and society in general when they are widely used. These new digital technologies play a complementary and catalytic role in the application of integrated technologies, presenting a Technology Route that is different from the elitist orientation.

3D printing is an additive manufacturing technology that uses physical, chemical and metallurgical techniques to create three-dimensional solid objects based on a three-dimensional digital model of a solid component

[57] This is the pithy summary of the vice president of an intelligent manufacturing company in Huangpu District, Guangzhou, which I researched in August 2021. The technology company featured above also advertises itself as a 'technology company that understands the process better.'

by continuously stacking materials layer by layer, incorporating cutting-edge technologies in the fields of digital modeling, electromechanical control, optoelectronic information and materials science. 3D printing currently faces the limitation of limited material variety (e.g., there are thousands of different types of metals available in traditional manufacturing, while in the field of metal 3D printing, only a few dozens of more mature raw materials are available); in addition, different processes have various unique requirements for material forms, while breakthroughs in materials science cannot be relied on existing intelligent and digital developments, but can only return to the study of the preparation and processing of materials themselves; here, therefore, digital and intelligent technologies have been adapted as indispensable aids, but do not define the entire operational process. Once breakthroughs in disciplines such as materials science have been achieved and are widely used, 3D printing can demonstrate the potential to revolutionize established ways of production and life: (1) by enabling design and manufacturing, management and production to be done in the same place, and by enhancing the capacity of companies to innovate on their own, it can further change the rigid global division of labor based on the internal division of labor within products, and advance post-Fordism at a new technological level.[58] (2) By being widely applied to all kinds of manufacturing products, it can simultaneously meet the batch production requirements of single, standardized products and the customized production requirements of diverse, individualized products, realizing. (3) By reducing the dependence on the size of the enterprise, it provides many opportunities for the start-up and development of small, medium and micro-enterprises.[59] It is because of the

[58] Post-Fordist production methods can help to reduce the over-dependence of enterprises on the global industrial chain, and most of the manufacturing enterprises in the United States that have successfully 'repatriated' earlier have post-Fordist 'flexible' 'team' production characteristics. For Chinese companies in the post-epidemic era, it is also important to explore production models that can better maintain the security of the industrial chain. For a case study of the return of US manufacturing firms, see Tim Hartsell and Dave Ripert, *The Road Back to US Manufacturing: Offshoring, Returning to the Homeland and Strategies for Revitalisation*, translated by He, Rong, Peking: People's Post and Telecommunications Publishing House, 2016, pp. 54–83.

[59] For a general introduction to 3D printing technology, see Du Yulei, ed. *3D Printing Materials*, Peking: Chemical Industry Press, 2020. On the socio-political significance of this technological development, see Cui, Zhiyuan, Dual Cycles, Global Value Chains and Additive Manufacturing: Preliminary Theoretical Reflections on New Development Strategies, *The 21st Century* (Institute of Chinese Culture, Chinese University of Hong Kong),

transformative potential of 3D printing that renowned business strategist Richard D'aveni sees a 'pan-Industrial Revolution' emerging: small-scale business platforms applying new manufacturing technologies are rapidly enabling some of the cutting-edge pioneers of large. The quality, speed and efficiency required by some of the leading pioneers to produce large volumes of a single product are rapidly being achieved, defeating the old business models based on economies of scale from which giant companies have long benefited.[60]

In addition, due to the current state of ambiguous data property rights, technology platforms with technological and capital advantages have easy access to rapidly expanding resources, creating a 'winner-takes-all effect.' In recent years, some institutional innovations have emerged to support the flow and sharing of data elements[61] and theoretical thinking.[62] However, the corresponding technical conditions are needed to support the effect of breaking the monopoly of data resources. In reality, there have been initial developments of new Technology Routes for Big Data to address this challenge, which deserves attention. For example, 'Social Linked Data' (Solid), a new Big Data Technology Route developed by Tim Berners-Lee, founder of the World Wide Web, in recent years, is centered on the sharing of data and information based on the will of individuals. This new technology enables users to store their data in freely chosen 'personal online data repositories' (pods); Solid-authenticated applications can request data after the user has granted permission to the application; and users can distribute personal information across several

August 2021 issue, no. 186 overall; see also Thomas Birtchnell, William Hoyle, *3D Printing for Development in the Global South: The 3D4D Challenge*, London: Palgrave Macmillan, 2014.

[60] D'Aveni, Richard, *The Pan-Industrial Revolution: How Manufacturing's Superheroes Are Changing the World?* translated by Ruxin Wang and Yanling Ye, Taipei: Aura Press, 2019.

[61] For example, the EU, the US, Singapore, South Korea and India have in recent years introduced regulations that establish or recognize the right to data portability, providing legal support to incentivize the free and orderly flow of personal data. This provision has the potential to break the monopoly of individual platforms over personal data. Article 45 of China's Personal Information Protection Law also comes close to this.

[62] For example, Zhang, Peng and Jiang, Yuhao, 'Basic theoretical research on asset-based management of government data: asset attributes, data ownership and pricing methods', *e-Government*, No. 9, 2020; Zhang, Huiping and Ma, Taiping, 'Cross-border flow of data elements in comprehensive digital transformation of cities: four models, promotion logic and innovation paths', *e-Government*, No. 5, 2022.

pods, both to ensure data security and to prevent a single developer from monopolizing the data.[63] Recently, it has been announced that Solid technology is being used in the UK to create an 'Open Bank' to promote competition in the banking sector. The new Technology Route for Big Data, presented by Solid, equips individual citizens with the technological tools to protect their data and enhance their autonomy.

Under the Common Wealth Strategy, China needs to mobilize its resources, both capital and professional, in line with the complex development issues it faces, and to more strongly encourage technological achievements that are conducive to building an inclusive economic growth model, serving to enhance the entrepreneurial and employment capacity of ordinary people and the development of small and medium-sized enterprises in various industries. The 14th Five-Year Plan for National Economic and Social Development and Vision 2035 outlines two critical measures, namely 'digital industrialization' and 'Digitalization of industry,' clearly demonstrating the policy idea of promoting the integration of digital technology and natural industry. Recently, a series of policy measures have been introduced, such as the Ministry of Industry and Information Technology and five other departments jointly formulated the 'Guidance on Promoting the High Quality Development of Light Industry,' requiring the implementation of digitalization to help build a modern light industry system.

At present, we can consider expanding the experience of the more process-savvy technology companies described above and the case studies of 3D printing and Solid and setting up some innovative mechanisms to gradually change the institutional environment for technology development and lead the adjustment of the Technology Route. For example, firstly, to set up a technology R&D platform for real manufacturing enterprises, to stimulate digital technology talents to go into the frontline of the manufacturing industry, to collaborate with skilled frontline workers, to develop new digital technologies that can be deeply integrated with various manufacturing technologies, to promote the formation of a new development model that breaks through the global Fordist division of labor and to create a new path for the growth of domestic small and medium-sized enterprises that are different from the original one restricted by the size of the enterprise and limited resources; secondly,

[63] For basic information visit the Solid website, https://solidproject.org/.

to innovate large technology companies and 3D printing and Solid case studies. Second, innovate the mechanism for the use of digital economy resources such as big data, guarantee the right of ordinary individuals to participate in the process of technological progress and economic and social development, and on this basis, further promote the development of digital technology and the creation of digital products closely around the specific needs of ordinary people's labor production and daily life and the accumulation of knowledge.

It is important to emphasize that the promotion of a new Technology Route will also require skills upgrading for workers and will expose enterprises to fierce competition and, therefore, oscillations and disruptions. However, it can avoid too drastic and disruptive changes by paying close attention to enhancing the capacity of workers' groups and MSMEs to participate on their own at every step of technological progress. The state can also pre-empt possible shocks and disruptions by putting in place the necessary innovative and protective systems and policies to help upgrade workers' skills and business operations and improve the ability of most people to defend themselves against risk. For example, supporting individuals and families in acquiring 'general skills for a broad and changing range of jobs,' providing more empowerment services for the general public and so on.[64] In any case, we advocate a new Technology Route, which opposes the separation of technological innovation from the manufacturing industry and the blind substitution of digital technology for human labor, and supports technological inventions and innovations that are more relevant to the actual development needs of the majority of ordinary people and therefore more viable. The state needs to encourage new technological products that empower ordinary people and improve social safety nets to support a highly innovative society.

Concluding Remarks

The biggest problem with the 'elitist' Technology Route is not its inability to provide innovative ideas and creative potential. However, its isolation from society at large disregards the productive practices and daily development needs of ordinary workers and other members of society, thus limiting innovative ideas and creative potential to a small number

[64] Charles Sabel, Dewey, Democracy, and Democratic Experimentalism, *Contemporary Pragmatism*, Vol. 9, No. 2 (December 2012), p. 43.

of people. It has become a factor that worsens the distribution of wealth and widens social divisions. 'The Technology Route initiative emphasizes the reorientation of new technologies so that technological progress is centered on the development needs and knowledge accumulation of the majority of workers and effectively serves the majority of ordinary people. In the face of the rapid development of a new generation of information and communication technologies, it is necessary to recall Proud hon's perceptive insight that science and technology should not be seen as a profession or a trade but 'as a competency that education should develop in every human being.'[65] This insight can be supported by researchers in the history of technological development, such as Lewis Mumford, a leading urban historian, who observed in the 1930s: 'Automated production has brought with it standardization, mass production and individual obscurity; ...... All the great advances of the machine system are based on manual labour and scientific thought. A machine system can never do more than the human eye, the human hand and the human brain that designed or operated it. With the knowledge of basic operations, man could theoretically rebuild any machine system in the world.'[66] ICTs have the potential to be more readily accessible and utilized by ordinary workers than the types of technologies that inspired the Fordist mass production methods during the Industrial Revolution and automation and therefore have a more decisive advantage in advancing flexible post-Fordist production methods. However, the development path jointly dominated by technological and capitalist elites has led to the exclusion of most workers and the knowledge accumulated daily by most real industries from technological innovation programs, making it difficult for the vanguardist promise that new technologies give to society to spread in a broader scale. This situation is unlikely to guarantee development prospects for inclusive growth.[67]

[65] Proudhon, *The Philosophy of Poverty*, vol. 2, translated by Yu. Shutong and Wang. Xuehua, Peking: The Commercial Press, 2010, p. 791.

[66] Lewis Mumford, *Technology and Civilization*, translated by Chen, Yunming, Wang, Keren and Li, Huashan, Peking: China Construction Industry Press, 2009, p. 368.

[67] For a critique of technological innovation as an exogenous variable in the practice of labor production, see Roberto M. Unger, The Knowledge Economy: A Critique of the Dominant View, *American Affair*, IV(3), Fall 2020. https://americanaffairsjournal.org/issue/fall-2020/.

The analysis in this chapter shows that countries can choose different Technology Routes in their development process to ensure that technological progress is not divorced from the needs of ordinary people, who make up the majority of the population. In order to develop and grow the digital economy in the new development context, we need to think clearly about digital technology development and application with the overriding principle of empowering different sectors and types of enterprises to develop and prevent the unreflective use of elitist Technology Routes to promote the digital economy for the sake of digitization, which could damage the industrial ecology and cause mass unemployment. Only by promoting a more significant number of information technology revolutionary achievements that are genuinely centered on the lives and productive realities of the majority of ordinary people and aimed at enhancing their labor productivity and entrepreneurial employ ability will it be possible to promote more sustainable innovation in the technology sector, while making digital technologies serve rather than hindering the creation of a 'development environment for all.'

CHAPTER 8

Change in Mechanism of Public Decision-Making Responsibility, Blockchain Technology and Break out of 'Command-and-Control'

In the era of big data, the volume, velocity, variety and veracity of data information[1] make it possible for public decision-makers to more closely track and serve the needs of ordinary citizens, but at the same time, it raises a variety of decision-making risks that are difficult to manage through traditional accountability mechanisms effectively. This chapter follows the 'problem-response' approach, first analyzing the challenges facing public policy-making in the era of big data and then discussing the possibility of changing the accountability mechanisms for public policy-making in the path of blockchain technology.

[1] Ward, Jonathan Stuart, and Adam Barker, Undefined By Data: A Survey of Big Data Definitions, https://arxiv.org/pdf/1309.5821.pdf.

Y. Jiang and K. Jia, *Transformation of Platform Governance in China*, IPP Studies in the Frontiers of China's Public Policy,
https://doi.org/10.1007/978-981-99-6456-7_8

The Challenges of Public Policy-Making in the Era of Big Data

(1) Decision-making risks arising from big data. Big data has been widely used in public governance areas such as urban transportation, urban planning and operations, public safety, social welfare and emergency management and has played a significant role in improving policy techniques, enhancing government capabilities and modernizing national governance systems.[2] However, it is worth noting that big data has a positive impact on public decision-making and poses risks.

A. While Big Data will enhance the science of decision-making, it will also lead to uncertainty. Big data has greatly expanded the boundaries of the information available to humans. However, the amount of data and information that people can process and know at this stage is very limited compared to the vast amount of data emerging at the basic level.[3] This has created a de facto 'cognitive paradox' in public policy-making: on the one hand, policy-makers have the certainty to implement policy design with greater precision and sophistication, but on the other hand, no policy-making process can claim to be exhaustive in terms of knowledge or algorithms. On the other hand, no decision-making process can claim to be exhaustive in terms of knowledge or algorithms. In other words, while Big Data provides a more reliable basis for scientific decision-making, it also increases the risk of uncertainty in decision-making—flaws in any of the elements of data quality, algorithms and models can have consequences contrary to the purpose of the decision. This increased risk of uncertainty creates correspondingly greater regulatory difficulties. In particular, with the widespread use of artificial intelligence technologies such as machine learning and deep learning, it is impossible for decision-makers to predict the subsequent behavior of artificial intelligence based on large-scale data acquisition. In this regard, James Kobielus, a prominent data engineer at IBM, points out the underlying reasons for the 'cognitive paradox': 'Big data-driven decision-making objectively requires automated models to capture as much of individual preferences or tastes

[2] Huang, Juan, and Huang, Zhuxiu, Big Data and Public Policy Research: Concepts, Relationships and Perspectives, *Chinese Administration*, 10, 25–30, 2015.

[3] Zhang, Nan, Public-Derived Big Data Analysis and Government Decision-Making Process Reconstruction: Theoretical Evolution and Research Outlook, *Chinese Administration*, 10, 2015.

as possible, resulting in increasingly complex decision models and increasingly complex decision processes. This has led to increasingly complex decision models and an increasingly opaque decision-making process. Any decision is attributed to any particular variable. Thus, while 'numbers' ultimately produce definitive policy recommendations, different outcomes driven by different numbers become more difficult to understand and interpret.'[4]

B. While enhancing the acceptability of decisions, Big Data will give decision-makers greater control. Big data-driven public policy-making is more relevant to the diverse and specific needs of ordinary people's lives than research studies that obtain 'small data' through sampling because extensive data analysis is more capable of revealing the multifaceted possibilities of life.[5] At the same time, big data is increasingly dependent on expert manipulation, from mining, collecting and managing to processing and utilization. At one end of the spectrum, this makes big data more closely linked to the everyday behavior and living conditions of ordinary people, helping decision-makers to gain a deeper understanding of their needs, while helping to make decisions more acceptable to ordinary citizens. However, at the other end of the spectrum, it increases the risk of decision-makers becoming detached from the feelings of ordinary people because of the threshold of expertise, knowledge and cost of use. In November 2014, a senior executive at Uber threatened to expose the private data of a journalist, sparking a lively intellectual debate on 'how to prevent data technologies from dominating human life,' given the ability of Uber's data to paint a panoramic picture of consumers' daily activities. This incident confirms the risk revealed by Dr Nicholas Diakopoulos, director of the University of Maryland's Computational Journalism Lab: 'Large data-driven algorithms will become the new power brokers in society, both in the corporate and government spheres.' [6] In other

[4] Kobielus, James, The Challenges of Transparent Accountability in Big Data Analytics, http://www.ibmbigdatahub.com/blog/challenges-transparent-accountability-big-data-analytics.

[5] Cukier, Kenneth, and Viktor Mayer-Schoenberger, The Rise of Big Data: How It's Changing the Way We Think About the World, *Foreign Affairs*, 92(3), 29–30, May / June 2013.

words,[6] data and the decision-making processes it drives do not necessarily lead to shared power and equality of status but instead may provide rulers or big business with more effective tools of control. Frank Pasquale, professor of law at the University of Maryland, pursues this question: any information we access online is recorded; the question is who has the right to access this data, for what purpose and for how long? Existing freedom of information and personal information protection laws are often at odds with state secrecy and trade secret laws and do not answer these key questions effectively.[7] In short, big data is relevant to everyone, but big data does not promise to be everyone's well-being.

C. Big data, while promoting complete openness as a basis for decision-making, will simultaneously hinder substantive public participation. The cost and technical requirements of using big data hinder its application by the general public, an issue that has attracted widespread academic attention. Jennifer Shkabatur, a fellow at the Kennedy School of Government at Harvard University, has pointed out that although the Open Government Memorandum signed by Barack Obama when he became President of the United States in 2009 initiated the development of open government data and 400,000 datasets had been put online by federal agencies by October 2012, empirical studies have shown that the public has had minimal access to these data.[8] Shkabatur's research suggests that while the era of big data presents a picture of open information and free access to data, the technology, equipment, cost and expertise required to use it has forced the public to rely more on big business, government and experts, thus preventing substantial public participation in the public decision-making process.

Related to the barriers formed by technology and knowledge, there are also many unresolved legal issues related to Big Data, thus creating institutional and legal barriers and hindering the general public's substantive participation in the use and development of Big Data. For example, what rights do individuals have over data relating to them? Can ordinary individuals use their private data as 'capital' to participate and share

[6] Diakopoulos, Nicholas, Rage Against the Algorithms, https://engenhonetwork.wordpress.com/2013/10/05/rage-against-the-algorithms-by-nicholas-diakopoulos/.

[7] Pasquale, Frank, *The Black Box Society: The Secret Algorithms That Control Money and Information*, Cambridge: Harvard University Press, 2006.

[8] Shkabatur, Jennifer, Transparency With(out) Accountability: Open Government in the United States, *Yale Law & Policy Review*, 31(1), 107–109, 2012.

developing and exploiting big data? The current general practice is that companies and governments with the technical and capital capabilities are the only users of Big Data. For example, in the pharmaceutical industry, it is common for pharmaceutical companies to enter into collaborative R&D agreements with individuals, where the former provides the technology, the latter provides their medical records, and the former hires the latter to conduct clinical trials. However, this collaborative model still limits substantive public participation in decision-making, and a more transformative approach would be to incentive patients to join together and use data as a basis for hiring companies to carry out collaborative research and development by safeguarding the rights of individuals to their private data.[9] To paraphrase political economy, it is not only necessary to encourage 'technology and capital to hire data' (to drive R&D) and break down institutional barriers to creatively enable 'data to hire technology and capital' (resulting in consumer-driven R&D). This will help ordinary people have more opportunities to use data to create projects that benefit them proactively.

D. While enabling polycentric decision-making models, Big Data will create difficulties in integration. In the era of Big Data, a polycentric pattern of public decision-making is created as governments, NGOs, enterprises and social groups create their own data networks, databases and cloud computing centers, and implement data mining, online analysis and exploitation operations related to their own business. However, as Charles Sabel and others have argued, a polycentric decision-making landscape requires a pluralistic and integrated mechanism to ensure that different stakeholders can work together to solve public problems.[10] Unfortunately, at this stage, there is no pluralistic integration mechanism around the use of big data, and three factors contribute to this problem:

Firstly, technical and institutional barriers to data interconnection still exist, such as data and system interfaces that have yet to be standardized, making it difficult to interface between various information systems.

[9] Evans, Barbara J., Barbarians at the Gate: Consumer-Driven Health Data Commons and the Transformation of Citizen Science, *American Journal of Law and Medicine*, 42(4), 2016.

[10] Sabel, Charles F., and William H. Simon, The Duty of Responsible Administration and the Problem of Police Accountability, *Yale Journal on Regulation*, 33(1), 165–211, 2016.

Secondly, the rights and responsibilities of the government, enterprises, society and individuals in the development of big data have yet to be clearly defined, while the structural problems within the government, as the promoter of the big data strategy, still need to be resolved. For example, in China, various departments and local governments have been reluctant to respond to requests for data disclosure, leading to 'supply promotion and demand suppression' in the area of open government data.[11] In the US, a similar situation exists in different forms, such as when the state of New York refused a federal directive to disclose illegal immigration data after Trump took office. Without commenting on specific incidents, we wish to highlight the complexity of government structures that constrain the formation of integrated public decision-making mechanisms. This potential risk will be magnified in the era of big data.

Thirdly, the exploration of data asset management mechanisms still needs to catch up lagging behind, making it difficult to reflect the value of data in the books, thus hindering further exploration of data exchange systems, updating mechanisms and improving of large-scale exploitation mechanisms. All in all, the fragmented state of data development has exacerbated the various divisions that exist in reality, which is not conducive to forming a pluralistic and integrated public decision-making pattern driven by big data.

It is important to note that the above-mentioned risks and challenges have yet to manifest themselves significantly in real life, as big data and its application in public decision-making is still in its infancy. However, they have already manifested themselves in some individual cases. For example, on the afternoon of 6 May 2010, the US Dow Jones Industrial Average fell by more than 1,000 points in minutes, losing US$1 trillion in market value in what was described as an 'instant crash.' According to the SEC's investigation report, published six months later, the sequence of events was such that there was a weak signal of falling prices in the market and all the high-frequency trading programs 'rationally' executed stop-loss decisions based on unique algorithms, closing positions at a frantic pace, leading to an unstoppable stock market crash. In this case, the electronic systems and data were not at fault. However, the data-driven uncertainty,

[11] Weng, Leyen, and Li, You-Yun, Openness and Sharing of Big Data in Government Affairs: A Study Of Conditions, Obstacles and Basic Guidelines, *Comparative Economic and Social Systems*, (2): 113–122, 2016.

the controlled and undemocratic nature of the back-office operations, and the fragmentation of multiple decisions geometrically amplified the danger signals and exacerbated the hidden decision risks.[12]

(2) The need for more traditional mechanisms of accountability for public decision-making. The public decision-making process is one in which decision-makers process the basis for their decisions within a given institutional environment. Big data provides a surprisingly more affluent basis for decision-making than the survey tools that have been constantly updated in the past, but how to ensure that decision-makers maintain their democratic accountability in this new environment is the question that needs to be answered. [13] Traditional accountability mechanisms for public decision-making do not facilitate decision-makers to address the risks of big data effectively.

The traditional accountability mechanism for public decision-making, based on a hierarchical system and emphasizing compliance with established rules, is by its nature both a vertical accountability mechanism (i.e., with top-down 'command-and-control' at its core) and a 'backward-looking' accountability mechanism (i.e., requiring 'present' decision-making to conform to 'past' regulations). The inappropriateness of this accountability mechanism in the context of Big Data is reflected in four main areas:

For one, because big data is constantly driving changes in decision-making, it is impossible to regulate decision-makers' behavior with established rules and procedures. Following the widespread application of concepts such as machine learning and deep learning to the big data industry (e.g., the rise of artificial intelligence), the ability of robots to deal with problems in unanticipated environments has become a reality, thus causing this uncertainty to become even more pronounced.

Secondly, because of the danger of expanding control powers hidden in Big Data, it is impossible to ensure decision-makers' public accountability by strengthening top-down oversight within the hierarchical system.

Thirdly, as the threshold of the cost of big data applications hinders substantive public participation, the protective provisions of existing regimes on the right to information, transparency and citizen participation may all be rendered ineffective in the face of technological or capital barriers.

[12] Kaplan, Jerry, *The Age of Artificial Intelligence*, translated by Li, Peng, Peking: CITIC Press, 2016, pp. 60–62.

Fourthly, the fact that multiple decision-making subjects exist also determines that the traditional top-down accountability mechanism based on a single subject no longer fits the needs of the times.

It is well known that the risks in the US sub-prime mortgage market in 2008 quickly turned into an unprecedented global financial crisis, precisely because a high level of uncertainty characterized the financial markets, a large number of participants, a long chain of risky transactions, shortened risk-holding periods and complex principal-agent relationships, which made it difficult for central decision-making bodies in each country to respond effectively. In a context where big data is challenging public decision-making, it is all the more essential to think about innovating public accountability mechanisms so that policy-makers are motivated and equipped to invest their wisdom in overcoming risks and ensuring the democratic nature of decision-making.

Data Information Governance and Accountability Mechanisms: A Review of Existing Research

A radical idea of change would be to call for the radical democratization of governmental hierarchies, as the famous futurist Alvin Toffler has proposed replacing hierarchies with ad hoc organizational forms so that 'the invisible geography of people and organizations will be replaced more and more rapidly, ... just as new nomads move from one place to another. nomads move from one place to another, so people move from one organizational structure to another day by day.'[13] In Toffler's view, only such ad hoc supranational forms could guarantee the free flow of data and information up and down the bureaucracy, inside and outside, and the flattening and democratizing of decision-making. However, such radical change is unlikely to occur in real life, as it has for Max Weber's social theorists and social scientists, since Max Weber's social theorists and political scientists have observed that 'bureaucracy, once established, is one of the most difficult of social organizations to destroy.'[14] Procedural and specialized hierarchical organizations have the

[13] Toffler, Alvin, *The Impact of the Future*, translated by Cai, Xinzhang, Peking: CITIC Press, 2006, p. 142.

[14] Weber, Max, *The Sociology of Domination*, translated by Kang Le and Jian Hhuimei, Guilin: Guangxi Normal University Press, 2004, p. 65.

function of reducing the cost of dealing with public affairs and are, therefore, an indispensable public institutional setup in modern societies.[15] However, Toffler's reflections have also helped researchers to recognize that decision-making institutions can be reconfigured with the help of emerging information technologies: by ensuring the sharing of data and information to widen the pathways for ordinary citizens to participate in public decision-making, they can contribute to a change in the logic of bureaucratic operations.[16] This understanding is the starting point for thinking about how to reconfigure the accountability mechanisms for public decision-making in the context of big data.

In practice, one of the more influential institutional innovations has been the use of open data information to strengthen democratic accountability in decision-making: mandatory disclosure of all data information that can be disclosed and holding decision-makers accountable to the public interest by exposing the basis and process of decision-making to the light of day. In January 2009, the US federal government introduced the Open Government Directive, promising 'unprecedented open, accountable government,' and in March of the same year, launched data.gov as a unified platform for unconditional data disclosure by federal government agencies and free public access to government information. In March of the same year, data.gov was launched as a unified platform for unconditional data disclosure by federal government agencies and free public access to government information, followed by a series of data disclosure acts, all of which are specific policy measures for data disclosure. However, there is room for reflection on both the theoretical underpinnings and the effectiveness of this influential policy approach, with public policy scholars pointing out, for example, that data disclosure is inherently a political game and does not 'naturally' reinforce the democratic accountability of public policy-makers.[17] Moreover, policy initiatives for data disclosure still need to overcome the aforementioned challenges of high costs, technical and intellectual barriers to using big data.

[15] Dorf, Michael C., After Bureaucracy, *The University of Chicago Law Review*, 2004, 71(3).

[16] Fountain, Jane E., Prospects for Improving the Regulatory Process Using: E-RULEMAKING, *Communications of the ACM. January*, 2003/Vol. 46. No. 1.

[17] Lord, Kristin M., *The Perils and Promise of Global Transparency: Why the Information Revolution May Not Lead to Security, Democracy, or Peace*, New York: State University of New York Press, 2006.

Two legal scholars address the shortcomings of whole disclosure policies: the need to go 'beyond disclosure.' They argue, based on numerous empirical cases, that full disclosure of data information using mandatory disclosure is not really beneficial: ordinary people do not bother to read lengthy 'informed consent' clauses and are less likely to learn a skill specifically because they care about a public decision. In a big data environment, attempts to increase the amount of information available to ordinary citizens to reinforce decision-makers' public accountability are unlikely to be effective. People often need 'advice' more than they need 'information.' The two professors suggest that we should go back to the traditional design of tort and contract law and re-emphasize the tort, breach of contract and reasonable faith responsibilities of producers, manufacturers, sellers, experts and others toward the public in order to urge the various social elites to abide by their duties on the one hand and monitor public decisions on the other.[18]

Contrary to thinking beyond disclosure, Joel Gurin's recent research on big data has re-emphasized the benefits of turning 'big data' into 'open data,' but unlike simply advocating for open data, Gurin points specifically to the incentives and advantages for big businesses with the technology and capital to use it. Gurin highlights the incentives and advantages of big businesses with the technical and capital capabilities to use big data.[19] Gurin's argument is supported by the institutional design of Archon Fung and others at the Kennedy School of Government at Harvard University: Fung proposes 'targeted disclosure' as an information rule, i.e., a requirement that specific types of information be specifically targeted to certain groups of people with application capabilities or special needs, in order to fully realize the social value of such important data or to assist specific subjects in making more effective use of helpful information.[20]

Other scholars have directly explored institutional innovations that enhance the ability of the general public to process data and information. For example, Cass Sunstein, a professor at Harvard Law School who served on the Obama White House staff, has proposed 'nudging,'

[18] Ben-Shahar, Omri, and Karl E. Schneider, *More Than You Wanted to Know: The Failure of Mandated Disclosure*, translated by Xiaofang Chen, Peking: Law Press, 2015.

[19] Gurin, Joe, Big Data and Open Data: How Open Will the Future Be, *A Journal of Law and Policy*, 10, 691–704 (2015).

[20] Fung, Archon, Mary Graham, and David Weil, *Full Disclosure: The Perils and Promise of Transparency*, Cambridge: Cambridge University Press, 2007.

arguing for policies that adopt 'a choice structure that changes people's behaviour in predictable ways, but does not prohibit any choices or change their economic incentives. To be called a nudge, the intervention must be something that can be easily and simply avoided. Boosting is not a command. Keeping fruit out of sight is an enabler, banning the purchase of junk food is not.'[21] In institutional practice, the concept of 'facilitation' is reflected in the requirement for governments to open up relevant meta-data and for open data to be machine-readable.

All of these research ideas are worthy of analysis. First, 'beyond disclosure' highlights the powerlessness of ordinary people in the face of big data and massive amounts of information and hopes to re-enforce the responsibility of government, enterprises and experts to make these technologies and capital-capable subjects effectively serve democratic decision-making; second, 'targeted disclosure' highlights the dominant role of technological and capital capabilities in the development of big data; and finally, 'facilitation' requires governments to take steps to enhance the cognitive capabilities of ordinary people to help them make rational judgments. This chapter argues that these studies highlight the point that the development of big data can only be achieved with the technological and capital capabilities of government and big business, but that public policy-making should never be based on technology and capital alone. Jonathan G. S. Koppell, an expert on network regulation, has argued that public accountability mechanisms in the network era should facilitate public decision-making to meet current social development needs and help ordinary citizens use network technologies to achieve their own development needs.[22] We need to consider: firstly, in terms of data and information governance, ensuring the credibility of data by forming a relationship of mutual supervision and mutual promotion among subjects with technical and capital capabilities, such as government, experts and large enterprises; secondly, in terms of responsibility mechanisms, establishing the responsibility of multiple subjects to jointly promote big data-driven public decision-making in the service of the public on the one hand, and forming the responsibility of helping ordinary people to participate as much as

[21] Thaler, Richard H., Cass R. Sunstein, *Nudge: Improving Decisions About Health, Wealth, and Happiness*, New Haven & London: Yale University Press, p. 6, 2008.

[22] Koppell, Jonathan G. S., Pathologies of Accountability: ICANN and the Challenge of 'Multiple Accountabilities Disorder', *Public Administration Review*, 65(1), 94–108, 2005.

possible in the decision-making process under big data on the other. On the other hand, the public responsibility to help ordinary people participate as much as possible in the decision-making process under big data should be established so that people who do not have the technical and capital capacity can learn by doing and continuously improve their ability to understand and apply big data.

How can this philosophy be implemented in concrete accountability mechanisms for public decision-making? This chapter discusses a new institutional innovation: using new networked information technologies to drive innovation in accountability mechanisms. Since its inception, the electronic currency Bitcoin has attracted a great deal of attention from online players and practitioners worldwide, and its underlying technology, blockchain, has been a major factor in creating a non-centralized, distributed, 'peer-to-peer' system. Its underlying technology, blockchain, has been highly sought after by corporate strategists, business strategists and public policy researchers for its creative creation of a non-centralized, distributed, consensus-based accountability mechanism with peer participation and oversight and has been credited with inspiring sweeping changes to new economic models and fundamental changes to government governance structures.[23] Blockchain technology can help to bring about a change in the accountability mechanisms of public decision-making with the concept above.

Blockchain Technology and Its Institutional Innovation Implications

In simple terms, a blockchain is a distributed ledger that allows traders to conduct peer-to-peer transactions directly and securely at any node without any intermediary or reliance on any over-arching monitor or credit guarantor. The entire blockchain comprises a series of 'blocks' containing transaction information, with the entire chain recording all information about an online transaction. Each block contains one or more transaction data sets, which are generated by random hashing (hashing) of the meta-data of the previous block and are automatically time-stamped. The newly generated hash, together with the transaction data for this

[23] Swan, Melanie, *Blockchain: Blueprint for a New Economy*, O'Reilly Media, 2015; Morabito, Vincenzo, *Business Innovation Through Blockchain: The B^3 Perspective*, Springer, 2017.

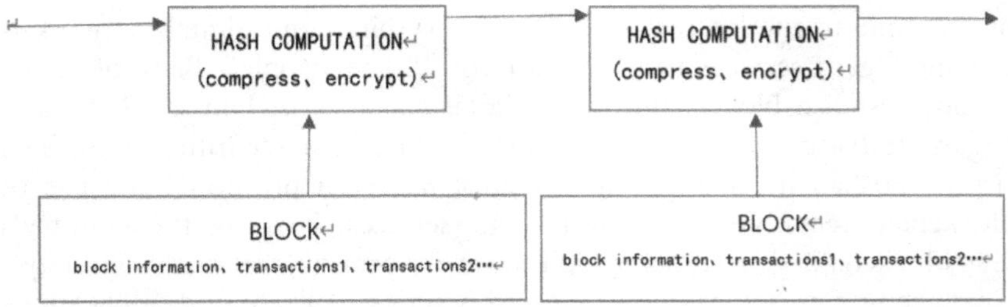

Fig. 8.1 Blockchain mechanism illustration

block, forms the meta-data for the next block, and the blocks are thus linked together to form a 'block chain.' Each block has a peer-to-peer access node and a panoramic interconnection of data (see Fig. 8.1).

The core feature of a blockchain formed based on the above mechanism is that all transaction information is stored on top of the chain, and tampering with any one piece of data must simultaneously modify the data stored in all blocks before the block in which that data is located, which is theoretically impossible to achieve; at the same time, any participant in the network can store the entire blockchain, thus forming a distributed storage where the collapse of any node will not affect the regular operation of the entire network. These two features—tamper-evident (or trace ability) and distributed (or peer-to-peer panoramic)—ensure that blockchain technology enables trust mechanisms under decentralization.

Of course, the reflections in this chapter do not suggest that blockchain technology has been perfected to the point where it can be used in any field. Indeed, hashing algorithms are constantly being revised and challenged. However, efforts to use the blockchain technology's security, efficiency and low-cost advantages to transform traditional financial services, commerce, cultural industries, government administration and other areas have been demonstrated in many countries and industries. Particularly noteworthy are recent developments in China: *the National Informatization Plan for the 13th Five-Year Plan* issued by the central government on 15 December 2016 included 'blockchain' for the first time in the development plan at the national level, stating that the current global informatization. The development situation is that 'new technologies such as the Internet of Things, cloud computing, big data, artificial intelligence, deep machine learning, blockchain and bio-genetic engineering are driving the evolution of cyberspace from the interconnection of everyone to the interconnection of everything, and digital, networked and intelligent services will be ubiquitous.' The People's Bank of China already tested a blockchain-based digital currency in January 2016 and began studying the implementation of blockchain securities rules. The city of Foshan in Guangdong Province has even promoted the use of blockchain technology in government services, creating the country's first blockchain government application innovation platform. The city's Chancheng District has now released a series of livelihood applications such as the Cloud Service Pass, the Pangu Health Platform and the Information Platform for the Elderly and Disabled. In short, mastering and applying blockchain technology in public governance is the need of the hour. China's exploration of institutional innovation from central to local levels fully illustrates the positive significance of this technology for institutional innovation in the public sector.

In the era of big data, public decision-makers should involve social actors with technical and capital capabilities to work together on public affairs, while taking responsibility for assisting ordinary people to participate in the development process. This chapter, therefore, suggests that blockchain, an information technology tool with a high degree of openness and mutuality, should be used to promote changes in the mechanisms of responsibility for public decision-making, ensuring that the capacity of public decision-making is significantly enhanced through information technology on the one hand, and overcoming the various risks posed by massive amounts of data on the other, strengthening the democratic nature of public decision-making and the responsibility of decision-makers to assist ordinary people in their development.

It is important to note that Bitcoin, the most significant successful blockchain application to date, is based on a political ideology and an operational mechanism that focuses on both requirements for responsible public decision-making. On the one hand, Bitcoin was born in the context of the 2008 global financial crisis and was designed to address the greedy and corrupt negative aspects of the modern financial system, bypassing the role of banks as intermediary structures in the modern financial system through a decentralized mechanism designed to enable direct peer-to-peer transactions and increase efficiency. In this way, Bitcoin, built on the blockchain, has in fact replaced, partially if not entirely, the traditional dualistic governance system of central banks and commercial banks through technological innovation, thereby dismantling the role of banks in the financial decision-making system and thus overcoming their decision-making risks. Regardless of the new types of risks that this 'substitution' may entail, it at least demonstrates the revolutionary potential of blockchain technology to transform accountability mechanisms for public decision-making. On the other hand, the decentralized nature of the blockchain does not mean that it will itself be in a state of 'anarchic' chaos; as Bitcoin has demonstrated for its nine-year history, the unique consensus mechanism allows each decision to be ultimately reflected in a unified public decision. The absence of top-down hierarchical authority makes this public decision fully accessible. The absence of top-down hierarchical authority makes such public decision-making fully democratic, which is also evidence of the innovative potential of blockchain technology to enhance the democratic nature of public decision-making. Section 4 of this chapter will further explore the paths and possibilities of blockchain technology to drive innovation in accountability mechanisms for public decision-making.

Blockchain Technology-Driven 'Command-and-Control' Breakthrough

Taking into account the technical characteristics of blockchain and the objective requirement of innovation in public decision-making accountability mechanisms, this chapter proposes the following five points to explore the possibility of changing the accountability mechanisms of public decision-making based on big data under the path of blockchain technology: breaking through 'command-and-control.'

Firstly, by leveraging blockchain technology, which is decentralized, non-tamperable and easily traceable, a mechanism of responsibility for public decision-making that is 'self-organized operation of government service platforms under government assistance' (collaboration) can be constructed. This requires a shift from compulsory open access to data and information to a structured responsibility for the data platform, monitoring its operation and handling disputes. For example, it requires that the data held by various entities—including government departments, social organizations, enterprises and private citizens—be hashed (with the help of third-party expertise) and recorded in the blockchain. Public institutions are only responsible for building, maintaining and overseeing the operation of the blockchain platform. Through blockchain technology, information about access and utilization of each data set is recorded and 'sealed' (i.e., it cannot be deleted or changed), thus ensuring security (as everyone can see who has accessed or added information) and protecting privacy (as information is de-tagged by a uniform algorithm), and thus facilitates the tracing and prosecution of others who use the data by individual subjects. Administrative and judicial mechanisms are only activated when a dispute arises between the subjects concerned, and decisions are made based on a well-developed data record on the blockchain. Blockchain technology thus allows government service platforms to operate effectively without centralized, top-down monitoring.

Secondly, the distributed nature of blockchain technology can be used to build a public accountability mechanism that 'facilitates macro policy-making and customized services' (linkage). With the integrity, detail and dynamism of the data recorded by blockchain, the peer-to-peer panoramic nature of blockchain technology can create a hierarchy between planned public policy-making and specific public service measures for different groups of people while ensuring the macroscopic nature of policy design and the targeting of specific service measures. For example, the objectives of welfare policies can be set based on macro-data. Then, when designing service measures, the government can change the practice of setting pre-defined standards for benefits or services and letting service recipients apply for them and instead provide personalized public service arrangements for different groups of people and individuals with the help of blockchain technology. As we all know, the overly regulatory approach of 'pre-defined criteria-application-approval' for public services generates a lot of auditing and monitoring work, leading to the expansion of public officials on the one hand and the need to prevent potential fraudulent applications on the other. The need to guard against potential fraudulent applications can also lead to a 'cat and mouse' relationship between the government and the people it serves. Blockchain technology ensures the traceability and authenticity of data information, which lays the foundation for customized public services, and due to the openness of the data information formed by blockchain, the government's investment in supervision can be significantly reduced, thus improving the efficiency and quality of public services.

Thirdly, by leveraging the flat participation characteristics of blockchain technology, a public accountability mechanism of 'flat decision making with the participation of multiple subjects' (sharing) is constructed. Blockchain technology gives different participating subjects the ability and motivation to maintain the security and openness of their data and information, and the relationship between multiple subjects to collaborate and check each other ensures that public decision-making does not need to penetrate into the details of social affairs overly, but only needs to make judgments in significant areas such as maintaining public safety, social order and development planning, thus forming a public accountability structure with formal and informal accountability mechanisms. At the same time, the public decision-making process based on the blockchain's open data is also supervised by various social actors, thus ensuring the openness of public policies.

Fourthly, by leveraging the collaborative openness of blockchain technology, a mechanism of accountability can be built that aims to push government agencies to 'look forward' (change the status quo). The openness and security of the blockchain provide an incentive for enterprises and other social actors with technical capabilities to fully exploit the commercial and technological value of data and information. At the same time, ordinary people can also enhance their own skills in participating in the development of big data by participating in the process of monitoring and tracking these data applications. In contrast, governments must take responsibility not only by following established rules but also by promoting the empowerment of ordinary citizens in future-oriented matters by strengthening the protection of data security, data ownership interests, privacy and orderly transactions, thus ensuring: the practical effect of urging governments, societies and ordinary citizens to govern collaboratively based on big data, so that public decisions are made in a way that promotes the achievement of the public interest The result is a practical effect that urges government, society and ordinary citizens to govern collaboratively based on big data so that public decision-making is based on the promotion of the public interest, and the strengthening of consensus in collaborative governance and the integration of polycentric decision-making through collaborative development.

Fifthly, to ensure the sound operation of blockchain platforms, the government's primary responsibility for creating public platforms (openness) must be strengthened. In order to promote the application of blockchain technology in the field of public decision-making, the government must assume some infrastructural responsibilities. Through the integrated construction of a 'general-to-general' and 'peer-to-peer' network system, on the one hand, the governance of people and goods will be fully integrated into the digital management system. On the other hand, the government has to overcome its own structural barriers to integrating and democratizing national certification capacity. Authentication refers to the establishment of a correspondence between data and people or things. As one of the basic national capabilities, the construction of authentication capacity will help to fully modernize the national governance system and governance capacity[24]; builds an open data information sharing platform; creates institutional ways and technical assistance for multiple subjects to participate in the opening and sharing of big data by the operational requirements of blockchain technology; and establishes rules for collaborative sharing and clear responsibility for data information to stimulate multiple subjects to support and promote the development of big data. The development of big data is encouraged by the formulation of rules for collaborative sharing and clear responsibility of data information. Strengthen the legal protection of blockchain. The blockchain, which is currently a self-organized form of network, relies mainly on mechanisms such as open-source intelligent protocols, cryptographic algorithms and codes to ensure its operation, and these mechanisms place too much emphasis on decentralized regulation and therefore carry the risk of instability. There is a need to establish a sound legal system compatible with big data and blockchain technology, with more rigid mechanisms to support the operation of self-organized platforms and enhance the protection of data security, privacy and orderly data transactions. Recently, European scholars have launched a study on how to apply blockchain technology under the EU General Data Protection Regulation (2016/679)[25] and promote the institutionalization of peer-review mechanisms. The government has promoted the formation of review panels by enterprises, social groups, experts and other government departments with big data processing capabilities so that the voluntary-based peer-review mechanism in blockchain technology can be institutionalized, forming a mechanism for comparison and mutual learning among various personalized public service measures, and supervising the continuous improvement of public services (see Fig. 8.2).

[24] Ou, Shujun, *The Foundation of National Basic Capacity*, Peking: China Social Science Press, 2013.

[25] Berberich, Matthias, and Malgorza Tastein, Blockchain Technology and the GDPR—How to Reconcile Privacy and Distributed Ledgers, *Eur. Data Prot. L. Rev*, 2, 2016.

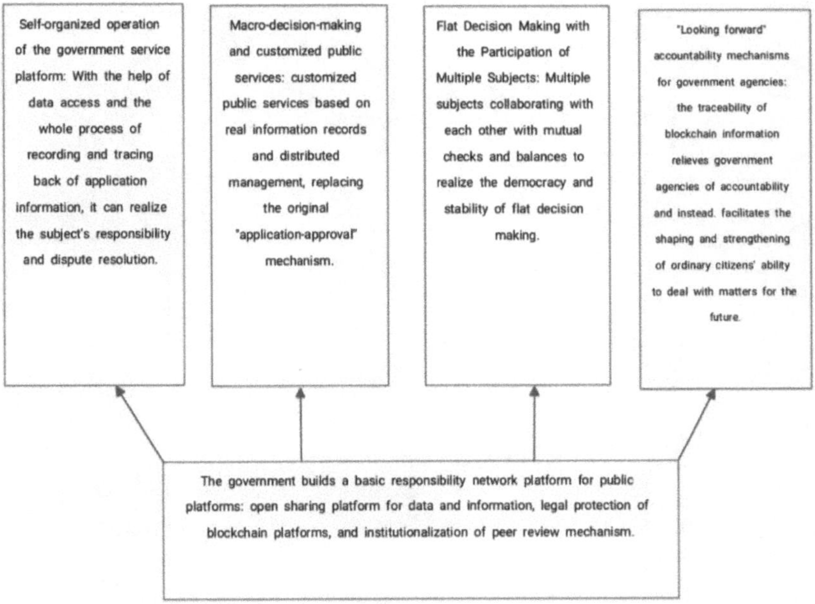

Fig. 8.2 Blockchain technology-driven innovation path for public decision-making accountability mechanisms

To sum up, blockchain technology can form a public accountability mechanism with multiple layers of collaboration (i.e., macro policy-making linked to customized services) and multiple sharing (i.e., government departments, enterprises and other social actors joined together), which can, to a considerable extent, facilitate public policy-making to overcome the risks of uncertainty, control, non-democratization and fragmentation brought about by the rise of big data. It is reasonable to expect that this new accountability mechanism will effectively ensure that public policies are made based on fully open and transparent big data, and that decision-makers are under the panoramic supervision of other relevant government departments, enterprises and society at large during the policy-making process, that they are more responsive to the needs of society and provide customized services to different groups and that they help ordinary people to have more opportunities to participate in decision-making and development.

CHAPTER 9

Construction of Responsible Government Through 'Collaboration-and-Sharing' Mechanism

The rapid development of the level of information technology, as well as the openness and transparency policies introduced by governments in recent years, has made open government data information—it is generally believed that data is the carrier of information and information is the value of data, but there is no clear-cut distinction between the two—has become the focus of social attention. However, whether open access to government data and information can enhance the ability of ordinary citizens to participate in and monitor public decision-making, and ensure that governments act more responsibly in the public interest, remains controversial.[1]

In January 2009, then-President Barack Obama signed the Memorandum on Transparency and Open Government, committing to an 'unprecedented level of open government' and a system of government linked to the three principles of transparency, public participation and collaboration. The federal government has taken several initiatives in response, such as enacting the Freedom of Information Act Memorandum to address the backlog of Freedom of Information cases, implementing the Open Government Directive, which requires government agencies to

[1] Erkkilä, Tero, *Government Transparency: Impacts and Unintended Consequences*, Hampshire: Palgrave Macmillan, 2012.

Y. Jiang and K. Jia, *Transformation of Platform Governance in China*, IPP Studies in the Frontiers of China's Public Policy, https://doi.org/10.1007/978-981-99-6456-7_9

release high-quality information proactively, and launching the Data.gov project to create a unified platform for storing government information based on the principle of 'open access.' In September 2011, the US, UK and eight other governments launched the Open Government Partnership in New York as a multilateral initiative to promote open government. However, despite these efforts, some empirical studies have pointed out that, firstly, it is often organized interest groups (such as businesses and social organizations) that have the incentive to invest resources and energy in actively accessing and utilizing massive amounts of data, and that opening up government information has not enhanced the ability of the general public to access and utilize big data; secondly, as the general public is kept out of the high cost of utilizing data and information, the effectiveness of formulating and amending relevant public policies based on public opinion is also limited. Secondly, as the general public is kept out of the high cost of using data and information, the effect of formulating and amending relevant public policies based on public opinions is not significant.[2] How to make open government information effectively available to different groups so that public policies can be more responsive to the needs of different groups remains a hot topic of academic research.[3]

China has been implementing the online government project since 1999, especially after the State Council promulgated regulations on government information disclosure in 2008, promoting the practical construction of e-government by central and local governments and their various departments and fully safeguarding the public's right to know has become one of the central government's priorities. On 1 July 2015, the General Office of the State Council issued the Opinions on the Use of Big Data to Strengthen Services and Supervision of Market Subjects (Guo Ban Fa [2015] No. 51), and on 15 December 2016, it announced the National Informatization Plan for the 13th Five-Year Plan (Guo Fa [2016] No. 73), which puts forward the modernization of the national governance system and the modernization of governance capacity, and put forward the requirement of using the means of great openness in government information to promote the construction of a responsible

[2] Shkabatur, Jennifer, Transparency With(out) Accountability: Open Government in the United States, *Yale Law & Policy Review*, 31(1), 79–140, 2012.

[3] Kosack, Stephen, and Archon Fung, Does Transparency Improve Governance? *Annual Review of Political Science*, 17(2), 65–87, 2014.

government. In a situation where the division of social interests is intensifying and public demands are growing, there is considerable urgency to explore in greater detail what kind of institutional construction can be used to ensure that the opening up of government information is used as a means to promote more transparent, rational and democratic public decision-making. This chapter proposes to discuss the path of institutional development from open government information to responsible government based on the existing research literature, taking into account the practical problems in the opening up of government data and information in China.

Accountability, Public Decision-Making and the New Governance: An Analytical Framework

The idea of a government accountable to the people has a long history in political thought, in both ancient and modern times. In the West, such a concept can be traced back to the ancient Greek and Roman idea of 'popular sovereignty.'[4] In ancient China, it was also said from a very early stage that 'Heaven reserves the people, and when the people want something, Heaven will obey it,' 'Heaven sees what the people see, and Heaven hears what the people hear' (Shang Shu—Tai Oath), and 'The people are born with a prime minister, and there are things and rules, and the people have their pinyin.' In ancient China, there were also early discussions of 'Heaven respects the people; if the people want something, Heaven will obey it,' 'Heaven sees what the people see, and Heaven hears what the people hear' (Shang Shu—Tai Oath), and 'The people are born with things and rules, and the people are good and virtuous' (Shi Jing—Cheng Min). However, while the understanding of accountable government has long been confined to the realm of electoral democracy and the possibility of accountability to the people only if a Western-style elected government is established, as Michael W. Dowdle has argued, different people have very different notions of 'public accountability.' These different understandings reflect different histories, experiences and

[4] Gordon, Scott, *Controlling the State: A Constitutional History from Ancient Athens to Today*, translated by Qi Ying et al., Nanjing: Jiangsu People's Publishing House, 2005, pp. 33–42.

problematic concerns.[5] The key is understanding the problems faced by accountability mechanisms in the context of public governance.

In a democratic society, the construction of a system of public accountability must face a threefold dilemma: (1) pluralism and fragmentation. From Jean-Jacques Rousseau's proposition that 'obedience to the 'public will' rather than to the 'will of the many',' it has been a fundamental principle that legislation and public decision-making should be based on the public interest rather than on the aggregation of individual interests. It has been a fundamental principle of the democratic ideal that legislation and public decisions should be based on the public interest rather than on the aggregation of individual interests. According to this principle, the people, rather than individual citizens or special interest groups, should be the subject of democratic accountability. However, the Western political system is based on the idea of individualism. The Western democratic system responds to a pluralistic and divided social structure with party politics, advocating that each interest group plays through its representatives in the legislative and public decision-making process, thus exacerbating social division and controversy, and to prevent one interest group from monopolizing political power, setting up a large number of 'veto points,' often leading to deadlocked policy negotiations.[6] Moreover, the pluralistic and polarized state of modern society is much higher than party politics presupposes: modern society presents itself as a fragmentation of opinions, rather than being based solely on static categories such as gender, race, class and status.[7] How to make public policies accountable to the people as a whole—rather than to voters or other special groups—becomes the first major challenge in the design of accountability mechanisms; (2) the division of political labor. The dual responsibility of professional politicians to express their opinions in response to public opinion on the one hand and to demonstrate their values in good faith on the other (e.g., by demonstrating their moral preferences and value positions when expressing public opinion)

[5] Dowdle, Michael W., Public Accountability: Conceptual, Historical, and Epistemic Mappings, in *Public Accountability: Designs, Dilemmas and Experiences*, ed. by Michael W. Dowdle, Cambridge: Cambridge University Press, 2006.

[6] Cameron, Charles M., *Veto Bargaining: Presidents and the Politics of Negative Power*, Cambridge: Cambridge University Press, 2006.

[7] Unger, Roberto M., *False Necessity: Anti-Necessitarian Social Theory in the Service of Radical Democracy*, London and New York: Verso, 2004, pp. 217–220.

often leads to political dilemmas for all kinds of officials. A case often discussed by political theorists is that of a Republican senator from North Dakota who voted against a bill in the US legislature in the 1990s to ban abortion, arguing that the legislation violated women's fundamental freedoms and was morally untenable, but who was himself from a conservative constituency known for being anti-abortion, leading to a debate about whether the senator had complied with or violated his political responsibilities.[8] In short, the design of democratic accountability mechanisms must also consider how politicians or decision-makers can balance maintaining their value judgments and respecting public opinion; and (3) the institutional basis for bureaucratic autonomy. In any larger system of governance, the bureaucratic system, designed according to the principles of the hierarchy, is an indispensable force. As early as the beginning of the last century, Max Weber saw the pessimistic prospect that a bureaucracy governed by purposive rationality would inevitably focus on the calculation of procedures, rules, ends and means in order to increase its operational efficiency, becoming an anti-democratic force that would maintain the secrecy of its knowledge and power, i.e., a closed 'bureaucratic identity group.'[9] There is, of course, surveillance of officials within the body politic and bureaucracy, but this surveillance, in turn, risks tying the hands of bureaucrats and thus shackling the bureaucracy's responsiveness to the needs of society.[10] It is worth noting that this chapter does not distinguish between the different forms of accountability of so-called politicians and bureaucrats, but rather uses the term 'government accountability' in general. As B. Dan Wood and Richard W. Waterman point out, from the 1980s onwards there has been a growing emphasis on the technical expertise of professional politicians, on the one hand, and the other, a trend toward greater political control of bureaucrats. These two changes have led to a convergence in the behavioral characteristics of

[8] Gutman, Ami, and Dennis Thompson, *Democracy and Disagreement*, translated by Lifeng Yang et al., Peking: Oriental Press, 2007, pp. 146–147, 154–156.

[9] Weber, Max, *The Sociology of Domination*, translated by Le Kang and Huimei Jian, Guilin: Guangxi Normal University Press, 2004, pp. 45–54.

[10] Anthony Downes identifies an important reason for the rigidity of action in bureaucratic organizations: the intermediate chain between decision-making and implementation is stretched as organizations grow larger, and regulatory rules have to be stacked on top of each other in order to prevent executive agencies from breaking the rules, thus creating a stifling of the ability to innovate. Downes, Anthony, *Inside the Bureaucracy*, translated by Xiaocong Guo et al., Peking: People's University of China Press, 2006, p. 169.

politicians and bureaucrats. It seems even more inappropriate to draw a strong distinction between the two in the context of the contemporary Chinese public policy-making system.[11] In short, there are three major challenges to designing accountability: how to ensure that the subject of accountability is not limited to a specific group of people with strong organizational and technical skills, but to the wider public; how to ensure that normative action is both responsive to the wishes of the people and a reasonable reflection of the professional opinion or value judgment of the decision-maker; and how to ensure that the basis for accountability often goes beyond considerations of organizational efficiency to respond to the actual needs of society?

With the rapid development of information technology such as the Internet, the scale and desire to exchange views have increased dramatically, and regulators face a far more complex situation than ever before. Jonathan G. S. Koppell proposes five critical dimensions of accountability design for a multi-regulatory system for the Internet: (1) transparency. Bureaucrats or other regulatory organizations must be accountable for their actions, subject to questioning, scrutiny and accountability; (2) accountability. The bureaucrat or organization can be identified as having a link to a specific consequence and the possibility exists that sanctions can be imposed for unaccountable behavior; (3) controllability. The control dimension is based on the traditional 'principal-agent' model of public administration: the bureaucracy or executive as agent must act at the behest of the people, the legislator or the superior officer as principal, not only to accomplish the intended task but also by procedures; (4) accountability. The most direct form of recourse is to judge whether the act is by the law, but because of the complexity of public affairs, the norms of public conduct should also remain somewhat flexible; (5) social responsiveness. The requirement for responsiveness can be divided into two different types: one is judged by the ability of the bureaucracy to respond positively to the 'needs' of the people; the other is judged by the ability of the bureaucracy to respond positively to the 'needs' of development. The other is based on the bureaucracy's responsiveness to the 'needs' of development. The former refers to a willingness for social autonomy, while the latter refers to a substantive goal set by the legislative or political elite (e.g., promoting equality, economic development and so on.). Of

[11] Dan Wood, B., and Richard W. Waterman, The Dynamics of Political Control of the Bureaucracy, *The American Political Science Review*, 85(3), 801–828, 1991.

the five dimensions, transparency is the most important, guaranteed and implemented within other accountability mechanisms.[12] Koppel's design is instructive in addressing the three main challenges mentioned above: transparency requires politicians and bureaucrats to explain the basis for decisions to the public by 'giving reasons in public,' so that decisions are not limited to the wishes of a particular group of people (e.g., voters or other interest groups), while accountability, controllability, recourse and social responsiveness establish the institutional conditions for officials to comply with the corresponding constraints while being given a certain amount of discretion to care for a broader public good or a more pressing particular interest.

However, Koppel's approach to accountability still needs to be expanded to the traditional regulation-accountability model, which sees regulation and accountability as the respective competencies of the two poles, the manager and the governed. In recent years, the emergence of direct citizen and frontline worker participation in public decision-making in complex areas such as information regulation, education, policing, ecological protection, health care and pharmaceutical management has led to significant changes in public governance systems in Europe and the US. This structural change, referred to by theorists as 'New Governance,' includes various innovations that have been described in the research literature as experimentalist governance, collaborative governance, networked governance and new public management. The essential feature of these innovations is to challenge the traditional governance model based on 'control-and-command' rules.[13] Although Koppel's vision is based on the decentralized regulatory mechanism of the Internet, it does not pay enough attention to the broader pluralistic participation, networking and dynamic regulatory features of the 'new governance' model.

According to the analysis of Charles F. Sabel and Jonathan Zeitlin, many institutional changes have taken place under the new governance model: for example, the duty of higher authorities to regulate the compliance behavior of frontline workers has been transformed into a duty to assist the latter in jointly exploring solutions, the mandatory obligation

[12] Koppell, Jonathan G. S., Pathologies of Accountability: ICANN and the Challenge of 'Multiple Accountabilities Disorder', *Public Administration Review*, 65(1), 94–108, 2005.

[13] See de Búrca, Gráinne, and Joanne Scott, eds., *New Governance and Constitutionalism in Europe and the US*, Oxford: Hart Publishing, 2006.

of officials to follow rules has given way to a dynamic responsibility to respond and solve problems, and the tendency of bureaucracies to be governed by purposive rationality has gradually opened up to the values of respect for participation and citizen autonomy. The tendency for bureaucracies to be governed by purposive rationality gradually increased toward respect for values such as participation and citizen autonomy.[14] Accordingly, the operation and meaning of accountability mechanisms have changed: firstly, the requirement for decision-makers to 'give reasons' for public policies is not just a passive presentation of relevant data and information but also an aid to citizens' understanding and use of such information. Secondly, public accountability cannot be designed under a 'principal-agent' model, as it is ultimately a result of the combined efforts of different actors, and it is difficult to attribute responsibility to a single entity. The accountability of officials has changed: instead of being held accountable for compliance with laws and regulations, officials are urged and assisted to deal with problems in advance.[15] It can be argued that public accountability becomes something like what Colin Scott calls 'extended accountability.'[16] Because of the complexity of the decision-making chain, it is not possible to establish a direct causal link between the act of making a decision and the consequences of the policy. However, it is possible for public responsibility can be set according to the effects of the multiplicity of actors involved in the decisions that public policy stimulates.

In other words, under the new governance model, a complex cycle is formed between accountability, public decision-making and open access to information, as shown in Fig. 9.1: the government opens up information about economic, socio-political data, which must have a substantial impact on the behavioral choices of ordinary citizens; as access to vast amounts of information is costly, the government and its experts should provide some support to assist citizens in understanding and using this

[14] See Sabel, Charles F., and Jonathan Zeitlin, Experimentalist Governance, in David Levi-Faur ed., *The Oxford Handbook of Governance*, Oxford: Oxford University Press, 2012.

[15] See Sabel, Charles F., and William H. Simon, The Duty of Responsible Administration and the Problem of Police Accountability, *Yale Journal on Regulation*, 33(1), 165–211, 2016.

[16] Scott, Colin, Accountability in the Regulatory State, *Journal of Law and Society*, 27(1), Voices, Spaces, and Processes in Constitutionalism (Mar. 2000), 38–60.

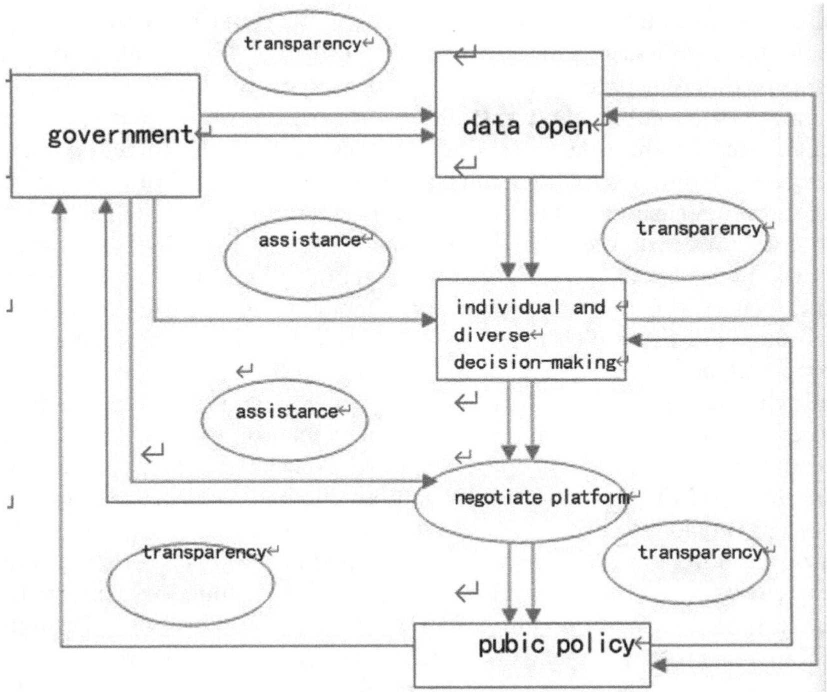

Fig. 9.1 Relationship between open data and the public decision-making cycle in the new governance model

data. The options developed by different individuals and collectives based on open data must be put together with the professional considerations of government for debate and consultation, ultimately resulting in public policy. The government has a responsibility to create such a consultation platform and to assist in providing professional counseling and data; data and information on the implementation of public policies should also be made available promptly through the open government data platform to help the public effectively monitor the implementation on the one hand and to empower multiple participants to make timely suggestions to amend policies and to urge improvements and refinements in decision-making on the other. In the whole decision-making process, the government's responsibility is not only to 'open up' government data but also to help individuals and collectives to understand the data, to help different groups use government data information for consultation and to urge different participants to open up their decision-making information.[17] This complex model of the relationship between open data and public decision-making more effectively responds to the threefold challenge of designing accountability mechanisms: first, governments must open up information about economic, social and political data that has a material impact on the behavioral choices of ordinary citizens, but because of the high cost of accessing vast amounts of information, governments and their experts should provide some support to assist citizens in understanding and using such data; second, different individual and collective proposals based on open data must be debated and negotiated alongside the professional considerations of the government, ultimately leading to public policy. The government has the responsibility to create such a consultation platform and to assist in providing professional counseling and data; thirdly, data and information on the implementation of public policies should also be made available promptly through an open government information platform to help the public effectively monitor the implementation on the one hand and to enable multiple participants to make timely suggestions for policy amendments on the other, to urge improvements and refinements in decision-making. In short, throughout the decision-making process, the government's responsibility lies not only in 'opening up' government information but also in helping individuals and collectives to understand the data, assisting different groups to use government data and information for consultation, and urging different participants to open up their decision-making information.

Data Information Barriers and Demand Repression: An Analysis of Related Structural Barriers

For a long time, the policy of opening up information on government data was not a consideration for China's ruling elite. Before the reform and opening up, the influence of the revolutionary tradition gave the

[17] Archon Fung and others offer a more straightforward but instructive model of the relationship between transparency policy and the public policy-making cycle. See Fung, Archon, Mary Graham, and David Weil, *Full Disclosure: The Perils and Promise of Transparency*, Cambridge: Cambridge University Press, 2007.

form of power organization in China a solid anti-rational bureaucratic dimension. As Jin Yaoji emphasizes, the revolutionary political and organizational philosophy dictated that the form of power organization at the time could not accommodate the idea of bureaucratic autonomy and the division of political labor.[18] Jin Yaoji points out the essential characteristics of the forms of power organization in China at the time: firstly, the anti-coordination system. The vertical hierarchy was rejected as giving rise to the stereotype of 'kicking the can down the road,' and the mobilization of the masses was the highest requirement for organizational activity. The necessary division of labor did not imply a separation of responsibilities, and the 'division of labor without division of labor' demanded that problems must be dealt with collaboratively; thirdly, the anti-specialist line. The 'red and professional' line was promoted, and the expert line was seen as an anti-political and anti-people proposition; fourthly, the anti-professionalization of the administration. Administrative leadership had to be integrated into the mass line, and the existence of a bureaucratic class that 'sat in the office and commanded' and 'talked without doing anything' was not allowed; fifth, anti-proceduralism. Administrative procedures were used as red tape to suppress the mass line, and the masses were encouraged to modify unreasonable organizational rules according to their wishes. The An Gang constitution of 'cadres in labor and workers in management' was the idealized form of the organization of power in this period, with two fundamental methods of organization: a system of role rotation and a system of collective decision-making. Needless to say, the denial of the political division of labor and bureaucratic autonomy, combined with the suppression of social pluralism, often made it difficult for state power to function correctly in the pre-reform period. Under this organizational philosophy of anti-rationalized bureaucracy, party cadres were more required to play a pioneering role than to assist in the autonomous development of society. Therefore, the institutional mechanism of political openness did not enter the vision of the political elite.

Since the reform and opening up of China, the country has witnessed a dramatic shift in the diversification of social interests. At the same time, significant progress has been made in the rationalization of the bureaucracy, and the responsibility of the public authority system to

[18] Yeo-chi King, Ambrose, A Voluntarist Model of Organization, The Maoist Version and Its Critique, *The British Journal of Sociology*, 28(3), 1977.

promote social development has been reaffirmed and emphasized in state policy. However, the institutional development necessary to deal with the complex social landscape still needs to catch up. The shortcomings in the area of government information or access to information are of two kinds:

First, serious data and information barriers have been created. The advantages and disadvantages of rebuilding China's bureaucracy coexist: although the hierarchy, laws, regulations and administrative procedures are gradually maturing, bureaucratic accountability is still based on a 'pressure-based system' consisting of technical means such as targets, projects, statistics and cadre appraisals,[19] and often leads to disadvantages: the stronger the accountability and the stricter the behavioral control, the more pronounced the tendency for bureaucrats to be accountable only to their superiors. The stronger the accountability and the more stringent the behavior, the more pronounced the tendency for bureaucrats to be accountable only to their superiors. This has led to a system of government that has grown in its ability to respond to problems but has kept its tendency to close off information and keep its operations secret. For example, around 2000, the promotion of e-government from the central to the local level aimed to improve the government's public service functions and responsibilities, but the construction of e-government around the world reinforced the disadvantage of data and information barriers between departments, resulting in 'strong vertical (within the departmental system) and weak horizontal (between departments).'[20] In January 2003, the Guangzhou Municipal Government Information Disclosure Regulations came into force as China's first local administrative law to systematically regulate the disclosure of government information, but in practice, it did not have the effect of overcoming departmental information barriers, instead relying heavily on the unsupervised discretion of departments in accessing information.[21] The Regulations of the People's Republic of China on the Disclosure of Government Information, a national administrative regulation formed on the basis of the experience of local governments in exploring the issue, is considered to

[19] Ran, Ran, Political Incentives and Local Environmental Governance in a 'Pressure-Based System', *Comparative Economic and Social Systems*, 2013(3).

[20] See Zhang, Weiying, and Liu He, eds., *Research Report on E-Government in China's Prefectural Cities*, Peking: Economic Science Press, 2003.

[21] Mo, Jiwu, Changming Yang, and Yuhao Jiang, *Consultative Democracy and Orderly Participation*, Peking: China Social Science Press, 2009, pp. 275–276.

have made significant progress in guaranteeing citizens' right to information, but scholars have also criticized the effect of its implementation as failing to break through the traditional sectional culture's hindrance to the disclosure of data and information.[22]

Secondly, it suppresses social demand. Although China's big data industry started late, the momentum of development is relatively rapid, not only many large enterprises or institutions such as Ali and Huitong World, which have a large amount of high-quality data as the source of social data commodities, but also comprehensive trading platforms such as Data Hall, Digital Ocean, Guizhou Big Data Exchange, Wuhan East Lake Big Data Exchange and Shanghai Big Data Exchange have been inaugurated in recent years; however, the current government data management system is hindering this development. China has implemented a 'multi-headed management and counterpart responsibility' system for managing government data and information resources: party and government departments at all levels and other enterprises and institutions are responsible for managing their data; many departments open and exchange data through business-related institutions. This multi-headed management system is based on an institutional framework of divided responsibilities, which is characterized by, firstly, the advantages of being highly professional and having the conditions to specialize in big data; secondly, it is inevitably subject to structural barriers such as multi-headed management and individual responsibility, which is not conducive to the opening and sharing of big data in government affairs; thirdly, due to the backwardness of laws and regulations and the severely restricted social supervision, it also makes it difficult for the various departments to carry out data opening and exchange. It is challenging to implement supervision on the actual situation of each department in fulfilling its responsibility of data opening and management. As some scholars have criticized, under such a management system, there is a strange phenomenon of 'construction on one hand and demand on the other, which ultimately leads to a disconnect between supply and demand': 'On the one hand, the Chinese government attaches great importance to information disclosure, setting up websites, opening institutions and allocating funds to support the relevant work. On the other hand, the government is doing so in a way

[22] Zhou, Hanhua, Exploring the Problems and Countermeasures of Implementing the Regulations on Open Government Information, *China Administration*, No. 7, 2009.

that discourages citizens' demand for information disclosure by setting thresholds and imposing additional conditions.'[23]

In short, China's current accountability mechanism is based on a 'pressure-based system,' and a division of departmental responsibilities, and the increasing tendency to rationalize the bureaucracy does not help to overcome the structural barriers that create data and information barriers and suppressed demand. The three main challenges of accountability design present themselves uniquely in contemporary China: first, the government's long-standing preference for its development planning is not yet responsive to the needs of a pluralistic society; second, the government's professional and technical capacity is focused on achieving policy goals and is far from adequate to help society's autonomous development needs; third, the bureaucracy is playing an increasingly prominent role in resource allocation, and organizational efficiency is rapidly increased. However, the constraints to prevent the bureaucracy from becoming detached from the needs of the general public have never been fully established. In general, since the reform and opening up of the country, the functioning of public power has gradually normalized, but the question of how to effectively respond to and promote the pluralistic needs of society is still open. As early as the late 1990s, scholars proposed a change from a pressure-based system to a cooperative one.[24] Such a change is only possible with fundamental changes in the accountability mechanisms of the bureaucracy.

In recent years, relevant national policies have vigorously advocated responding to diverse social needs while promoting the industrial development needs. On 5 *September* 2015, the State Council issued the Notice on the Issuance of the Action Plan for Promoting the Development of Big Data, proposing the concept of 'accelerating the construction of a strong data country' and 'vigorously promoting the interconnection and open sharing of government information systems and public data, accelerating the integration of government information platforms, eliminating information silos, promoting the opening of data resources to society, enhancing the credibility of the government, guiding social development

[23] Li, Weng, and You-Yun Li, Openness and Sharing of Big Data in Government Affairs: A Study of Conditions, Obstacles and Basic Guidelines, *Comparative Economic and Social Systems*, (2), 118, 2016.

[24] See Jingben, Rong, Cui Zhiyuan et al., *The Shift from a Pressure-Based System to a Democratic and Cooperative System*, Peking: Central Compilation Press, 1998.

and serving public enterprises.' The Interim Measures for the Management of Government Information Resources Sharing, promulgated on 5 *September* 2016, stipulate that open sharing is 'demand' oriented and requires 'providing information technology to the public.' In the 2017 Central Government Work Report, it was reiterated that 'the interconnection of information systems of State Council departments and local governments should be accelerated to form a nationwide unified government service platform.' In the 2017 central government work report, the government reform was reiterated as 'accelerating the interconnection of information systems of the State Council departments and local governments, and forming a national unified government service platform.'[25] In particular, it is worth pointing out that, unlike in 2008, when the Central Government introduced the Regulations on the Disclosure of Government Information, the new round of opening up of government information was implemented in the context of the 'Internet+' and the policy of mass entrepreneurship and innovation, and local governments were able to understand in greater depth the economic and social value of government information and thus showed greater inclusive. Local governments have shown more enthusiasm in promoting this work (e.g., a large number of local innovations have emerged: Peking's unified open platform and government data information catalogue management system, Shanghai's open data application innovation competition, Guizhou's big data industry development, Wuhan and Lanzhou's smart city construction and so on). Some scholars have said that open government in the context of 'Internet+' has a different meaning and can only be the subject of reform if it is placed on the same level of importance as political ideas such as transparency and democracy.[26] With national and local policies showing strong momentum, it is time to think about the redesign of public accountability mechanisms in the context of open government information: Through the redesign of accountability mechanisms, the bureaucracy can be pushed to break through the pressure system and

[25] Report on the Work of the Government—5 March 2017 at the Fifth Session of the Twelfth National People's Congress, http://lianghui.people.com.cn/2017/n1/2017/0316/c410899-29150065.html.

[26] See Jia, Kai, From 'Open Source Software' to 'Open Government': The Change of Government Governance Under the Influence of the Internet, *Comparative Economic and Social Systems*, (2), 2016.

division of departmental responsibilities, laying the institutional foundation for sustainable openness and sharing of government information on the one hand, and nurturing the government's motivation and capacity to assist in the development of social diversity on the other. On the other hand, the government's motivation and capacity to contribute to developing a diverse society will be nurtured.

Promoting Democratic Accountability Under the 'Collaboration-and-Sharing' Mechanism

The social theorist Emile Durkheim has said that freedom of contract can only be guaranteed on a non-contractual basis. The political scientist Stephen Holmes has also argued that democratic societies should be supplemented by non-democratic institutional mechanisms to ensure that free and equal but differently endowed and otherwise qualified individuals can develop fully in society.[27] The system of government and bureaucracy cannot become an open democratic organization that breaks away from formal rationality, hierarchy and administrative conventions. However, it can, through the innovation of accountability mechanisms, become an indispensable and fundamental institution for the proper functioning of a democratic society.

In contemporary Chinese society, it is a fact that plural interests are highly polarized. In practice, based on the premise that pluralistic demands are on the rise, institutional innovations have emerged in public governance to explore pluralistic participation and networked modes of governance.[28] Accordingly, the role and function of the bureaucracy must change: from leading social development in a dictatorial manner to balancing the need to promote development with pluralistic demands; from focusing on serving policy goals with professional competence to assisting pluralistic subjects in negotiating shared governance; and from unilaterally maintaining its organizational efficiency to serving diverse needs. In his 2016 Government Work Report, Premier Li Keqiang said he would 'vigorously promote 'Internet + government services' and realize

[27] See Holmes, Stephen, Lineages of the Rule of Law, in *Democracy and the Rule of Law*, eds. by Jose Maria Maravall and Adam Przeworski, Cambridge: Cambridge University Press, 2003.

[28] See Yijia, Jing, ed., *The Road to Collaborative Governance in China*, New York: Palgrave Macmillan, 2015.

data sharing between departments, so that residents and businesses can run fewer errands, do better work and avoid congestion. Simplify the removal of hassles and harshness and prohibit the investigation of illegality, so that the people have more equal opportunities and more room for creativity.' In the 2017 Report on the Work of the Government, it was once again emphasized that the community can enjoy the effectiveness of the reform of the 'government's own revolution,' which is mainly composed of the 'management and service': 'We must let enterprises and the public feel more We must let enterprises and the masses feel more the effectiveness of the reform of 'management and service,' make efforts to open up the 'last mile,' and resolutely get rid of troublesome disadvantages, implement fair measures and open the door of convenience.' In this new development, there is a need to think specifically about institutional innovations to overcome the disadvantages of China's bureaucratic system that lead to data information barriers and suppress social needs, and to create a path where 'the government leads the reform, serves a plurality of social subjects, and promotes the people to share the results of the reform': by consciously assisting and motivating the public to share information about government affairs, the government will be able to cultivate the public's interest in government affairs. By consciously assisting and incentivizing the public to share data and information on government affairs, we can cultivate multiple beneficiaries of the collaborative sharing of government data and information, create pressure for public accountability, and promote and improve the construction of a responsible government.

In terms of public accountability mechanisms, three ideas for institution building can be proposed as follows:

One to set up and improve institutional mechanisms for government decisions to be accountable to the public interest: the bureaucracy recognizes and respects the fact that society is pluralistically divided but facilitates the formation of social consensus by creating platforms for openness and consultation of different interests. Firstly, it must be clear that major public decisions should be based on open data and information, and that public decisions in general should at least provide explanations and justifications of the basis for the decisions so that the public can easily question them through various means; secondly, in the process of opening up data and information, the government should establish mechanisms to facilitate access to and use of government information by a plurality of social actors, thereby fostering a plurality of actors who can hold the government to account. For example, the traditional data or information disclosure policy represented by the US Freedom of Information Act only emphasizes the disclosure of corresponding data by the public sector. However, it fails to focus on the use of data. These include setting standards for the machine-readability of data, opening up data on government decision-making processes, improving the visualization and accessibility of online services and creating online channels for the public to propose technological innovations to the authorities.[29] Through these mechanisms, a vital link is established between 'institutional supply' and 'social demand' to ensure that the opening up of government data and information is sustained long term and effectively.

Secondly, it is essential to set up and improve mechanisms that allow the government to respond to people's demands while at the same time creating a platform for rational consultation: the bureaucracy and government experts set issues based on people's demands and help to raise the level of awareness and communication. The need for government to respond to people's demands is fundamental to the design of modern accountability mechanisms, and the responsibility of politicians to prevent people from being biased by information asymmetries or imperfect knowledge systems is a central requirement of modern open government systems. Bruce Ackerman and James S. Fishkin of Yale Law School have argued that, in order to ensure that voters make rational choices in general elections, it is necessary to set up a 'consultation day' a week or so before the election, during which citizens are exposed to as much information as possible about candidates and political issues, as well as to the public. This would allow citizens to be exposed to as much information as possible about the candidates and political issues, to engage in a whole dialogue, to consult and exchange views on the election and then to reach a new understanding and so on.[30] This idea is, of course, still only a thought experiment, but the problems it points out and the ideas it suggests deserve attention: on major public issues, responsible governments and politicians should collate and fully open up data and information according to what the people need to know, to help them reason—this is what makes the 'division of labor' in a democratic society

[29] See Gascó-Hernández, Mila, ed., *Open Government: Opportunities and Challenges for Public Governance*, New York: Springer, 2014.

[30] Ackerman, Bruce, and James S. Fishkin, Deliberation Day, *The Journal of Political Philosophy*, 10(2), 129–152, 2002.

possible. This is the basis for properly functioning the 'division of labor' in a democracy. Interestingly, Fishkin found this method of government and its grassroots officials listening to and assisting people's voices in a democracy forum in the town of Zeguo in China's Zhejiang province: township officials released information on many township infrastructure proposals, which were explained in detail to the villagers, who then negotiated and discussed the implementation (including proposing changes) of some of these proposals.[31]

Thirdly, setting up and improving mechanisms for the government to assist people in using information from government data to participate in public decision-making: by providing opportunities to learn and access a broader range of data, the bureaucracy uses its specialized skills to assist people in participating in public decision-making. A very well-known example in this regard is the participatory budgeting system in Porto Alegre, Brazil: each year, a budget committee is formed by a certain number of representatives elected by the residents' assembly to discuss the preparation of the municipal budget for the following year; prior to the formal budget discussion, the government arranges a 'learning session' for the representatives to learn about the public budget and provides data and other technical support at the time of the budget committee to help the delegates fully express the budgetary needs of their community.[32] In this institutional practice, the technical expertise of bureaucrats and politicians is valued without being divorced from the needs of the population; data and information are made available for the autonomous development of the community, without the differences in technical skills making large amounts of or specialized data inaccessible to the general public.

Of course, these three ideas are still a long way from concrete institutional design, and the challenge is to urge governments to respond positively and effectively to the increasingly diverse needs of society. The Interim Measures on the Management of Government Information Resources Sharing promulgated by China's State Council in 2016, which aims to 'regulate the sharing of government information resources between government departments,' have taken a big step forward in

[31] Fishkin, James S., *When the People Speak: Deliberative Democracy and Public Consultation*, Oxford: Oxford University Press, 2009, pp. 106–110.

[32] See Baiocchi, Gianpaolo, Participation, Activism, and Politics: The Porto Alegre, in Deepening Democracy, *Institutional Innovations in Empowered Participatory Governance*, ed. by Archon Fung and Erik Olin Wright, London and New York: Verso, 2003.

breaking down sectoral barriers. However, there is still much to be explored in terms of how to ensure that the opening up of government data is based on social needs, and how to urge the bureaucracy to assist ordinary citizens and other social actors in implementing their right to participate in public decision-making.

Concluding Remarks

In any case, the 'opening up' of government data and information is only the first step in constructing a responsible government. The opening up of government data and information not only involves the activities of the government's work responsibilities but is also an opportunity to present, guide and strengthen the diverse needs of society. In the new era of development, this must be deeply understood. The 19th National Congress of the Party redefined the principal contradiction in Chinese society as 'the contradiction between the people's growing need for a better life and unbalanced and inadequate development' and specifically pointed out that the people have 'growing demands in terms of democracy, rule of law, fairness, justice, security, environment, etc.' The main theme of the new era is to lead the people to fully participate in the reform process and share the fruits of reform through institutional innovation. Against the backdrop of China's central government vigorously promoting the opening up of data and information to achieve a revolution in the government itself while promoting a policy of mass innovation and entrepreneurship, there is an urgent need to explore the establishment of a collaborative sharing system for government data and information, so that the general public can truly share the value of government information by fostering and assisting multiple subjects to use government data and information to make public and private decisions. Accordingly, redesigning the public accountability mechanism, for example, creating accountability pressure on the bureaucracy to assist the general public in realizing their development needs, 'pushing' governments at all levels to revise unreasonable rules and regulations actively, and changing the bureaucracy's logic of upward accountability and division of responsibilities between departments, is a measure worth trying. This is how to move toward a new model of governance that is more open, democratic and accountable.

CHAPTER 10

Innovation in Management Mechanism of Big Data for Government Affairs: From 'Command-and-Control' to 'Collaboration-and-Sharing'

On 5 *September* 2015, the State Council issued the Notice on the Issuance of the Action Plan for Promoting the Development of Big Data (Guo Fa [2015] No. 50), setting the development goals of forming a cross-sectoral data resource sharing pattern by the end of 2017 and building a national data unification and opening platform by the end of 2018. On 7 January 2016, the National Development and Reform Commission issued *the Notice on Organizing and Implementing Major Projects to Promote the Development of Big Data* (Development and Reform Office High Technology [2016] No. 42), setting out the development goals of 'carrying out the application of big data for public services,' 'establishing and improving. The policy level has been vigorously promoted, creating a new environment for the development of big data. The vigorous promotion at the policy level has created the primary conditions for the opening and sharing big data for government services in China.'

However, the management of big data for government affairs in China is still in its early stages, and there are many structural barriers, such as lagging legal norms, sectoral barriers and a disconnect between supply and

Y. Jiang and K. Jia, *Transformation of Platform Governance in China*, IPP Studies in the Frontiers of China's Public Policy,
https://doi.org/10.1007/978-981-99-6456-7_10

demand.[1] There are also contradictions and conflicts between the data rights and demands of multiple subjects. It is worth paying attention to and researching how to guide multiple subjects to jointly participate in the construction of big data for government affairs through the innovation of data management mechanisms.

Concepts, Types and Management Structure Characteristics of Big Data in Government

Big Data for government affairs is data and information generated or used by party and government departments and other enterprises and institutions with public management functions in the course of performing their functions, including data related to the state of existence of government departments, data generated in the course of the operation of government departments, data formed after collection, processing or conversion in the course of the implementation of management by government departments and so on.

There are four primary sources of government data: the first is data formed by departments and agencies in the course of carrying out their statutory functions, called 'business data,' which refers to data collected and generated in the course of business operations; the second is public opinion and social data, which refers to data obtained by government departments through statistical surveys of individual social enterprises; and the third is environmental data, which refers to data obtained through physical equipment such as meteorological, environmental and visual data; the third category is environmental data, i.e., meteorological, environmental and imaging data obtained through physical equipment collection.[2] In addition to these three categories, data that exists in a decentralized form in the hands of social subjects is also becoming increasingly important. In particular, in recent years, social capital has invested in establishing a large number of research institutions, corporate research institutes and data development organizations, which hold a large amount

[1] Weng, Leyen and Li, You-Yun, Openness and sharing of big data in government affairs: a study of conditions, obstacles and basic guidelines, *Comparative Economic and Social Systems*, 2016(2).

[2] Mu Yong, Top-level design framework and application scheme for Peking's big data platform for government affairs, Available from: http://www.haokoo.com/internet/1840119.html.

Table 10.1 Types of data resources for government services

Serial number	*Species*	*Application status*
1	Operational data	Generally structured by government departments and used to a broader extent
2	Public Opinion Data	Generally structured by government departments and used to a broader extent
3	Environmental data	Less structured or semi-structured, the range of applications is widening
4	Decentralized public data	Usually structured by companies and institutions, with a wide range of social applications

of data relevant to government-public decision-making. This kind of data can be called 'decentralized public data,' and the government can obtain the right to use it for public decision-making purposes through various means, such as government procurement or cooperative development. The four data types differ in terms of their degree of structure and scope of application (see Table 10.1).

The most distinctive feature of the management structure of big data in the Chinese government is that it is 'multi-headed management with counterpart responsibility'; that is, central and local party and government departments and other public institutions are responsible for the management of the data they hold, while within the same institution, there is usually a special information management department; accounting, auditing and data asset value assessment institutions are involved in the management process, taking on some of the data management responsibilities depending on their functions. Different data users are responsible for different stages of the data operation and maintenance process. The advantage of this multi-headed management structure is that it is highly professional and can make full use of the professional conditions of each department to specialize in the processing and management of big data; its disadvantage is that it relies too much on the discretion of departments, creating structural barriers such as departmental barriers, which, on the one hand, is not conducive to the open sharing of big data in government affairs and, on the other hand, makes it challenging to implement supervision of the actual situation of each department in fulfilling its data management responsibilities. In 2015, the State Council's Action Plan for Promoting the Development of Big Data put forward the instruction of 'accelerating the integration of government information platforms and eliminating information silos,' and in the same year, the Office of the Leading Group for Network Security and Informatization of the Central Committee of the Communist Party of China issued the Opinions on Strengthening the Network Security Management of Cloud Computing Services in Party and Government Departments (Zhongguo Netifafawen, 2015). In the same year, the Office of the Leading Group of the Central Committee of the Communist Party of China for Network Security and Informatization issued the 'Opinions on Strengthening Network Security Management of Cloud Computing Services in Party and Government Departments' (Zhongguancun Fafawen [2015] No. 14), which strictly 'enhances risk awareness and responsibility awareness,' precisely to address such shortcomings in data management structure.

In contemporary society, the extent and scope of the use of big data in government are expanding rapidly. Previously, the management of government data tended to emphasize the creation of large unified databases for data aggregation and management. However, in recent years, there has been a growing trend toward the use of decentralized outlets for data collection, which has resulted in the inability to create a unified and centralized management platform and database. For example, the US Food and Drug Administration Amendments Act of 2007 requires that medical data at the scale of 100 million or more people is used for post-marketing safety performance monitoring studies of pharmaceutical products. The FDA established surveillance sentinels and their branch outlets for this purpose, collecting data on 60 million people by 2011. Similar large-scale pharmaceutical surveillance studies have been conducted in Canada, the European Union and Japan. The collection and use of such massive amounts of data can be done without requiring or making it possible to build centralized databases to manage them.[3] By way of comparison, China's decentralized, multi-headed management structure has advantages worth continuing to explore. However, there is an urgent need to explore innovations in data management mechanisms to enable the decentralized multi-headed management structure to function correctly under the requirements of unified openness and collaborative sharing for the vigorous promotion of open sharing of government data.

[3] Barbara J. Evans, Much ADO About Data Ownership, *Harvard Journal of Law and Technology*, 25(1), 2011, pp. 99–100.

The Data Titling Process and Its Shortcomings

Decentralized multi-headed management structures face two significant problems that prevent open sharing and make it challenging to enforce oversight, a similar situation in the US and European countries. One popular proposition is to treat data and information as property and to promote open access through a property rights management approach that balances the rights of data owners.[4] This proposition is quite influential and needs to be specifically identified. Theoretically, there are two different approaches to data property rights: 'private property right' and 'public property rights,' and in the US, the debate between the two approaches has focused on the specific issue of healthcare data resource management. The debate on the two different approaches to property rights in the United States focuses on healthcare data resource management.

(1) The path to private property rights for data resources. In the US, the issue of ownership of medical data held by the executive branch and the databases of medical institutions falls within the jurisdiction of state authority. However, the states have not enacted laws on related issues. For many years, patients' medical data has been protected by the privacy provisions of the Health Insurance Portability and Accountability Act (enacted in 2002), and there are many federal policies and ordinances regulating research activities involving human subjects, giving strict protection of personal data. However, for researchers, this static protection hindered scientific research by preventing the sharing of medical data resources and made it difficult for patients to access treatment. As a result, around 2009, many commentators and state legislators began calling for legislation to facilitate the open sharing of medical data.

However, in practice, there are many subjects, such as doctors, healthcare providers and insurance companies, who hold information on patient data in addition to the patients themselves. Who has the right to decide to what extent the data they hold can be used?

Mark A. Hall argues for the privatization of data resources: the ownership of medical data belongs to the patient. Accordingly, whether, how and to what extent medical data should be made available (for a fee or free of charge) is a matter for the patient to consider on his or her own

[4] Mel Hogan and Tamara Shepherd, Information Ownership and Materiality in an Age of Big Data Surveillance, *Journal of Information Policy*, 5, 2015.

merits. At the same time, in order to stimulate joint discussion and scientific research on the patient's condition, medical and scientific institutions can be authorized to openly use such data on the basis of the individual's consent. Hall specifically states that the patient's 'consent' need not be 'express,' as in the case of market transactions, based on equal negotiation, but need only demonstrate that the patient has reasonable access to the purposes for which research and medical institutions will use the data. The patient's 'consent' is presumed. Of course, medical and scientific institutions must 'anonymize' patient data when using and opening it up.[5]

(2) A public ownership approach to data resources. Marc A. Rodwin proposes a public property rights approach to data management: since healthcare data resources are vital to the overall well-being of society, the ownership of healthcare data cannot be considered as belonging to any 'private subjects' such as patients, healthcare providers and insurance companies, but should be a public property right. It should be a public property right. The various actors should hand over the medical data they hold to a specially established public body, which would judge whether to open it up and to what extent, based on public interest objectives such as scientific research—for example, unrestricted access for non-profit research institutions and fees based on costs and expected benefits for profit-making pharmaceutical companies. Implement fee-based access based on costs and expected benefits. Similarly, public bodies must anonymize data when opening up healthcare data.[6]

The most critical aspect of Lowe's view is his vision of a public body for data management. He argues that this public body, which would manage data ownership on behalf of the public, should be 'trusted' and 'qualified.' Specifically, this public body should have three conditions: firstly, a strict internal management code that can be subject to rigorous auditing, with the ability to prevent abuses by managers; secondly, a well-developed technical expertise to ensure professional processing and management of the health data collected; and thirdly, a complete openness system that can be opened up by law and on request.

[5] Mark A. Hall, Property, Privacy and the Pursuit of Integrated Electronic Medical Records, *IOWA Law Review*, 95, 2010.

[6] Marc A. Rodwin, Patient Data: Property, Privacy & the Public Interest, *American Journal of Law & Medicine*, 36, 2010.

(3) The shortcomings of the data titling process. These ideas of data titling are not limited to medical data but have also emerged in discussions of innovative data management mechanisms in many other areas and have had an impact on legislative efforts, for example, in the State of Texas in the United States, where state legislators have proposed data titling legislation in 2010. In its extreme form, the right to privacy can also be seen as a property right, reflecting exclusive and exclusive domination of data resources. This chapter argues that the data titling approach, while influential, has serious flaws and does not accommodate the fundamental characteristics of big data:

The private property rights approach is too biased toward protecting the rights of patients and does not consider the social value of medical data. A study by an American scholar of the collection of medical data by scientific institutions found that the data held by these institutions, when based solely on individual patient consent, created a statistical 'bias': it was often patients with common medical conditions who voluntarily opened up their data. In contrast, patients with more critical abnormalities for scientific work were generally reluctant to open up their medical data. Patients with medical conditions that are more important for research are generally reluctant to release their medical data, thus leading to an inherent bias in the data collected based on patient consent.[7] Predetermined sampling is a traditional data collection method, which is incompatible with the nature of so-called big data.

The path to public ownership of data faces the problem of being too costly and prone to 'inefficiency.' In modern societies, the increasing refinement of the professional division of labor and the enormous amount of data available make it impossible for public institutions of the kind envisaged by Lowy to emerge. For example, the standardized management and anonymization of large amounts of data is a task that can hardly be carried out by any single institution on its own, and it is also almost impossible to require a public institution to have a full complement of skilled personnel from all disciplines in order to specialize in the processing of data, given the many disciplines of modern medicine.

Further, both approaches to data property rights imply a presupposition of 'absolute property rights,' which is incompatible with the characteristics of big data. Absolute property rights mean that a certain

[7] Barbara J. Evans, Much ADO About Data Ownership, *Harvard Journal of Law and Technology*, 25(1), 2011, p. 76.

subject has the exclusive right to certain social resources. However, in the case of big data, firstly, the process of data collection and processing and open use involves multiple subjects whose activities often intersect with each other and are not easily distinguishable; secondly, the open use of data is a process of constantly adding materials and investing resources, and the data resources or information products eventually formed are the result of multi-step operation. In other words, the so-called big data result from the superposition of multiple subjects in multiple operational steps, so it is not appropriate to grant any single subject (whether private or public) exclusive property rights and interests. For example, it is unclear whether the information in a medical record is the personal data of the patient, the professional data of the doctor performing the diagnostic activity or the professional data of the medical institution. In reality, the progress of scientific research has been hampered by overly strict property rights protection. For example, the US patent law that grants patent protection to the discoverer of a gene sequence, resulting in restrictions on subsequent inventions, has been strongly criticized by the scientific community.

Under the assumption of 'absolute property rights,' the mechanism of data management is necessarily dominated by a 'command-and-control' mechanism: absolute property rights in data necessarily presuppose that a single subject of rights—whether private or public—has control over the data (in the terminology of property economics, the exclusive enjoyment of 'residual claims'). The idea of absolute property right to data necessarily presupposes that a single subject of rights—whether private or public—has control over the data (in property economics terms, the exclusive enjoyment of 'residual claim') and that the top-down 'command-and-control' mechanism is. The only means of exercising this right/power is through top-down 'command-and-control' mechanisms.

In short, there are significant obstacles to the open data sharing under the data property rights approach, and one cannot innovate China's government data management mechanisms based on this thinking.

Toward 'Collaboration-and-Sharing': Innovative Mechanisms for Managing Big Data in Government Affairs

(1) Ideas for innovative data management mechanisms. At present, governments around the world have made many valuable attempts to socialize the use of government data in areas such as education, tourism, transportation, population and business registration information, and a model has emerged in which the government unifies the establishment of platforms, and various business departments participate together.[8] The Interim Measures for the Management of Government Information Resources Sharing, issued by the State Council in 2016, also stipulate that each government department is responsible for linking its department to the data sharing and exchange platform and implementing openness and sharing through that platform (Article 4). This is a practical management approach that has been explored in practice. As noted earlier, the key here remains how to address the two significant challenges of openness barriers and difficulty in implementing oversight.

As analyzed in this chapter, the property rights approach is insufficient to address these two challenges, as it inherently implies disadvantages that prevent open sharing. There is a need to find another solution: to break away from the overly rigid question of 'who is entitled to exclusive rights and interests in data' and instead innovate the data management mechanism to ensure that different actors can benefit from the open sharing of big data. Research in public management reveals that combined legal, administrative and market-based mechanisms can create a pattern of joint participation of multiple subjects in public governance.[9] The opening up big data for government affairs involves the interests of multiple subjects, such as the central government, local authorities, different departments, other public institutions, ordinary citizens, enterprises and other social groups. Therefore, it is precisely necessary to establish a data management mechanism with clear responsibilities and collaborative sharing so that different subjects have both the pressure to implement openness by the law and the opportunity to share the benefits of opening up big data,

[8] Peking opens up 'big data' of government information resources for social use, *People's Daily*, 2013-7-22.

[9] Charles F. Sabel and William H. Simon, Contextualizing Regimes: Institutionalization as a Response to the Limits of Interpretation and Policy Engineering, *Michigan Law Review*, 110(7), May 2012.

thus forming a joint effort to promote the sustainable opening and use of big data in government affairs.

(2) The theoretical basis for innovative data management mechanisms. To deny absolute property rights is not to deny rights holders their fundamental rights and interests. Studies in the economics of law have already pointed out that property rights are not absolute but rather a 'bundle of rights,' which includes many different rights such as occupation, use and income, and can be broken down and given to different subjects so that they can realize their interests without interfering with the enjoyment of their rights by others. In 1972, Guido Calabresi, in collaboration with A. Douglas Melamed, published 'Property rules, liability rules and prohibited transactions rules: a perspective on the cathedral of Rouen,' which provided valuable theoretical insights for further institutional design.[10]

In brief, Calabresi and others argue that the protection of rights can be discussed in terms of entitlement. Firstly, the law determines whether the subject of the right can freely exchange certain rights. For example, citizens' rights to freedom of belief, human dignity and so on cannot be freely exchanged. These rights and interests are protected by the 'no-trade rule.' In the context of big data, data closely related to personal privacy, state secrets and public security should also be protected by the 'no-trade rule.' 'This aspect is not discussed in this chapter.' Secondly, and more importantly, among the rights and interests that the law allows for free exchange between rights subjects, they can be protected by two different legal protection methods, the 'property rule' and the 'liability rule,' the legal effect of which is that the rights and interests protected by the property rule can only be exchanged with the consent of the subject of the right. In contrast, an interest protected by the liability rule can be compulsorily exchanged without the consent of the subject of the right, by giving him reasonable compensation. As shown in Table 10.2, within the binary framework of 'empowerment' and 'legal protection methods,' the design of Calabresi et al. avoids granting absolute property rights to one subject alone but rather guarantees the relevant rights and interests of different subjects.

This chapter argues that the advantages of the design mentioned above solution lie in the fact that, through the combined use of property and

[10] Guido Calabresi and A. Douglas Melamed, Property Rules, Liability Rules, and Inalienability: One View of The Cathedral, *Harvard Law Review*, 85, April 1972.

liability rules, mechanisms can be created with different levels of protection for social resources with different characteristics. Firstly, the property rules (i.e., Rules 1 and 3) have a strict protection function but are not conducive to promoting cooperation and are therefore suitable for social resources with clear property rights and clear ownership, which can be put to good use even without the use of social exploitation; secondly, the protection approach of the liability rules (i.e., Rules 2 and 4) emphasizes the relativist nature of rights and interests, and is more appropriate for resources with more excellent social benefits; thirdly and, most importantly, legislators can develop different protection measures for different social resources based on a combination of factors such as social value and public interest. In the case of big government data, for the part of government data that is more controversial in terms of ownership or social benefits, the protection measures are designed according to the liability rules, which can, on the one hand, promote the flow and utilization of government data resources and realize the social value of the data, and on the other hand, set up reasonable compensation means to compensate for the losses of the relevant rights and interests. Moreover, according to the liability rules, although there is the possibility of implementing 'compulsory exchange,' at the same time, the subject of the compulsion is allowed the opportunity to appeal. It is guaranteed to receive appropriate compensation, which does not create the risk of infringing on the legitimate rights and interests of the compelled person.

(3) 'Three, two, four' type of big data management mechanism for government affairs. This chapter attempts to innovate the management mechanism of big data for government affairs by combining property and liability rules. The innovation aims is to establish institutional ways for multiple subjects to participate in the opening and sharing of big data, and to stimulate multiple subjects to support and promote the development of big data in government affairs through collaborative sharing and clear protection rules of responsibility. This chapter designs a 'three-two-four' data management mechanism, including three basic principles, two exchange platforms and four protection rules. Functionally speaking, this data management mechanism fully respects and builds on the current decentralized multi-headed management structure while clarifying the openness responsibilities and sharing obligations of different government data departments, and on the other hand, takes into account the value of decentralized public data held by enterprises and other social actors, and therefore establishes a smoother data exchange channel between the government, society and the market. The specific management mechanism is designed as follows:

Table 10.2 Different protection of interests under different rules

Empowerment	*Legal protection methods*	
	Property rules	*Responsibility rules*
Residents	**Rule 1** (giving residents the right to be protected from pollution and not to pollute them without their consent)	**Rule 2** (gives residents the right to be free from pollution, however, even without their consent, a third party may judge whether to create a certain level of pollution by giving them compensation)
Companies	**Rule 3** (the rights and interests granted to the enterprise to develop its production and not to interfere with its normal production activities for any reason without its consent)	**Rule 4** (giving the enterprise the right to develop its production, however, even without the consent of the enterprise, a third party may judge whether to stop its production activities by granting compensation)

A. The three basic principles.

The first basic principle: the ownership of government data resources is owned by the entire population. The process of their generation, and operation and maintenance management relies primarily on the support of the public purse.

The second basic principle: government departments, enterprises and institutions undertaking public administration functions are responsible for the management of government data formed within the scope of their statutory duties and activities and for opening up the data on a unified open government data platform by the law.

The third basic principle: the opening up and use of government data resources must be aimed at promoting social benefits, enhancing the public decision-making capacity of party and government departments, supporting commercial activities with significant social value or helping citizens to exercise their legitimate political and economic rights more effectively.

B. Two exchange platforms.

The first platform is establishing a unified open platform based on the principle of free and open access to government data resources.

The second platform: for non-free and open government data, establish a consultation platform for data exchange between the government and society and enterprises, including establishing institutional means to obtain redress in case of unsuccessful consultation.

The first platform falls under the protection of Table 10.3, 'Rule 1'; the negotiated part of the second platform falls under the protection of 'Rule 3,' and the unnegotiated part falls under the protection of 'Rule 2' and 'Rule 4.'

C. Four rules of protection.

This is a vital part of the 'three-two-four' data management mechanism (see Table 10.3 for detailed design). Rule 1 is the embodiment and concrete implementation of the first and second basic principles, ensuring that government data is 'open as the norm'; Rule 2 is the remedy to Rule 1 and the institutional way to judge whether the government has reasonable reasons for not opening up data, ensuring that enterprises, social groups and ordinary citizens have sufficient access to government data; Rules 3 and 4 establish the institutional way for the government to obtain data from social subjects, and implement the third fundamental principle in the exchange process; Rule 4 establishes an institutional passageway for the government to obtain data from social subjects and implements the third fundamental principle in the exchange process.

The comprehensive application of the four rules creates conditions for the government, citizens, enterprises and other social groups to share the value of big government data and realize their respective data rights and interests. It clarifies the responsibilities of multiple subjects in opening and sharing to ensure the social value and sustainable development of big government data. It can be expected that protecting the benefits of open government data shared by different subjects can create pressure and motivation to overcome the disadvantages of open barriers while clarifying the open responsibilities of different subjects can also create institutional conditions for promoting and monitoring the continuous opening of government data. Through the protection of legitimate rights and interests and setting responsibilities for openness, the 'three-two-four' data management mechanism aims to lay the foundation for sustainable openness and sharing of big government data.

Table 10.3 Big data protection rules for government services

Protection rules	*Content of the rules*	*Basis for rule design*
Rule 1	In general, the ownership of government data resources is owned by the whole population and the government opens them to the community free of charge according to the purpose of public interest. For data of exceptionally high economic and social value and high technical requirements for exploitation, corresponding conditions can also be set to open it up for specific public or enterprises. Products developed based on targeted open data can only be priced by the developer (research institutions, enterprises or other service providers) based on the development costs and limited revenues invested.	Existing laws and administrative regulations have confirmed the public nature of government data resources. For example, Article 11 of the Law of the People's Republic of China on Surveying and Mapping stipulates that 'basic surveying and mapping are public welfare undertakings'; Article 14 stipulates that 'People's governments at or above the county level shall incorporate basic surveying and mapping into their annual plans for national economic and social development and financial budgets.' Another example is Article 27 of the Regulations of the People's Republic of China on the Disclosure of Government Information, stipulating: 'Administrative organs shall not charge fees for providing government information upon application, except for the costs of retrieval, reproduction and postage. Administrative organs shall not provide government information through other organizations or individuals by way of paid services'

Protection rules	*Content of the rules*	*Basis for rule design*
Rule 2	If a citizen, enterprise or other social group believes that a specific type of government data resource has more significant social and economic benefits, but the government does not open it up, the citizen, enterprise or other social group can apply to a higher government or the court to request the government to open it up. Citizens, enterprises or other social groups must pay the corresponding consideration: (1) in the case of exploitation and reuse for profit, a consideration based on the market benefits of the data product; (2) in the case of exploitation and reuse for non-profit purposes, a price determined based on the cost of producing the data. The consideration paid by social agents becomes the benchmark price for the relevant data product on the market.	The Regulations of the People's Republic of China on the Disclosure of Government Information stipulate the scope of information disclosure and the relief mechanism for the administrative counterpart. For example, Article 33(1) of the Regulations provides that 'Where a citizen, legal person or other organization believes that an administrative organ is not fulfilling its obligation to disclose government information in accordance with the law, it may report the matter to a higher administrative organ, a supervisory organ or the department in charge of government information disclosure work. The organ receiving the report shall investigate and deal with it.' Paragraph 2 provides that 'If a citizen, legal person or other organization considers that the specific administrative act of an administrative organ in the work of government information disclosure has infringed upon its lawful rights and interests, it may apply for administrative reconsideration or file an administrative lawsuit in accordance with the law'

(continued)

Table 10.3 (continued)

Protection rules	*Content of the rules*	*Basis for rule design*
Rule 3	Where data resources collected and developed by citizens, enterprises or other social groups in the course of their daily operations are only auxiliary to (i.e., have no direct role in) the government's performance of public administration functions, the government must acquire the right to use them using an equally negotiated transaction. The government may expropriate decentralized public data that is significant to the public administration function by the law, but reasonable compensation should be paid.	For example, Article 31 of the Law of the People's Republic of China on Surveying and Mapping stipulates that 'Basic surveying and mapping results and other surveying and mapping results completed with state investment shall be provided free of charge if they are used for decision-making by state organs and for socially beneficial purposes. Those outside the provisions of the preceding paragraph shall be subject to a system of paid use in accordance with the law; however, the government and its relevant departments and the military may use them without compensation for the needs of public interests such as disaster prevention, disaster mitigation and national defence construction'
Rule 4	Where data resources collected and developed by citizens, enterprises or other social groups in the course of their day-to-day operations are complementary to the government's performance of public administration functions, the government may appeal to the courts to obtain the right to use such data by paying a price based on the cost of development plus a limited advance price, if the parties fail to negotiate a deal.	Ibid

CHAPTER 11

Asset-Based Management of Government Data: Design in the Context of the 'Collaboration-and-Sharing' Mechanism

With the deep integration of the new round of information and communication technology and economic and social development, the use of big data and other technical means to enhance public governance has become an integral and essential part of the national strategy, and the construction of systems in related fields is rapidly being put on the agenda. For example, the Fourth Plenary Session of the 19th CCP Central Committee called for 'improving the mechanism whereby factors of production such as labour, capital, land, knowledge, technology, management and data are evaluated by the market for their contribution and remunerated according to their contribution.' The Opinions of the Central Committee of the Communist Party of China and the State Council on Building a Perfect Institutional Mechanism for Market-Based Allocation of Factors, published on 20 March 2020, proposed to 'accelerate the cultivation of the market for data elements' and identified three specific measures to promote the open sharing of government data, enhance the value of social data resources and strengthen the integration and security protection of data resources. The Civil Code of the People's Republic of China, adopted at the third session of the 13th National People's Congress, also responds to new developments and issues in the new ICT era in the

Y. Jiang and K. Jia, *Transformation of Platform Governance in China*, IPP Studies in the Frontiers of China's Public Policy,
https://doi.org/10.1007/978-981-99-6456-7_11

form of statutory law, providing either specific or principled protection for personal information (Article 111, Articles 104 to 109) and data (Article 127) through a series of explicit provisions. China has entered the fast lane in advancing institutional innovation in the digital age.

It should be acknowledged that the open application of government data resources and the management of big government data in China are still at an early stage of development. There is much room for improvement at the institutional and policy levels, such as the governance of big government data, the incentive compatibility of multiple departments, the construction of development and utilization mechanisms and trading platforms, the coordination of asset-based management methods in the financial and accounting systems, and information security. There are many issues that need to be studied in depth, such as the protection of personal information. Among them, the asset-based management of government data resources could be a strong area in terms of both practical and theoretical discussions. Topics such as determining the asset attributes of government data resources, the principles of data ownership and the pricing methods that can be adopted have not been answered by researchers so far. Based on the analysis of the current situation, this chapter intends to discuss the academic literature at home and abroad and provide preliminary thoughts on some fundamental issues of asset-based management of government data, to arouse the general concern of the academic and practical communities.

Basic Concepts of Asset-Based Management of Government Data

(1) **'Government data.'** There is no unified and specific formulation or definition of government data, and there are various terms such as government data, public data and government information in academia, government, and industry, and the scope of the definition inconsistent. This chapter adopts a relatively concise 'working definition': in terms of content, government data (also known as government data resources) refers to a category of significant data resources related to government business areas, including data related to the state of existence of government, data generated in the course of government operations, data generated in the course of government operations and implementation management, and data generated in the course of government

operations and implementation management. The data collected, produced or transformed in the course of government operations and implementation management and so on.

In line with what is known in practice, and in terms of the current work of Chinese party and government departments, enterprises and institutions with public administration functions, there are four main types of government data sources and the principles of open access for different types of data are not the same, nor are their asset pricing methods (the pricing methods for different data will be discussed later):

The first category is the data formed while performing statutory functions by various government departments and agencies, which can be called 'government business data,' referring to the data that the government has the right to collect and generate during its business. For example, data in production and construction, data on urban infrastructure, data on transportation facilities, data on resources, taxation and finance, data on public services such as health care and education.

The second category is 'social and public affairs data,' which refers to data obtained by government departments on their initiative or by commissioning other organizations to conduct statistical surveys on social enterprises and individual subjects. For example, census data, economic census data, financial supervision data, anti-money laundering monitoring data, food and drug supervision data and so on.

The third category is 'basic environmental data,' i.e., meteorological, environmental and imagery data acquired through physical equipment. In China, the authority to collect this type of data is also vested in the government. However, the law provides that society may use it for public interest purposes without compensation, as provided for in Article 31 of the Law of the People's Republic of China on Surveying and Mapping.

The fourth category is 'decentralized public data,' i.e., a large number of scientific research institutions, enterprise research institutes, data-open enterprises and so on, which are invested by social capital and hold a large amount of massive data related to the government's functions and the implementation of public decisions. At present, with the development of the big data industry, the government, enterprises and other subjects have a large amount of data resources. However, due to the lack of an effective data sharing and exchange mechanism, the phenomenon of 'data islands' is particularly significant. There is a need to consider legislative and other regulatory means to enable the government to access such decentralized public data through compulsory acquisition, procurement and cooperative development to meet public decision-making needs.

(2) **'Asset-based management of government affairs data.'** The rapid development of cloud computing, artificial intelligence and other new-generation information and communication technologies has made big data an increasingly popular theme of the times, and the use of big data as an effective tool to support government decision-making on significant issues, promote government management reform and enhance government services has become an inevitable choice to promote the modernization of the national governance system and governance capacity. Accordingly, exploring various forms of data opening and realizing the value-added use of government data resources through multiple channels has become a strategic point in the new era. In the era of big data, the economic and social values hidden in the data collected, produced or transformed by the government and various departments, agencies and social organizations in the course of performing their functions need to be fully explored.

In 2011, a European Commission assessment stated that opening up and reusing public sector data information could be worth €40 billion per year to the EU economy; in the same year, Deloitte, one of the Big Four accounting firms, assessed that the direct economic benefit of opening up public information in the UK was £1.8 billion per year.[1] A 2014 mckinsey report stated that G20 governments' open data strategies would lead to an increase in output of US$13 trillion over the next five years, which would be equivalent to driving 1.1 percentage points of GDP growth in G20 countries. Based on this understanding, the Ministry of Industry and Information Technology's 2017 Big Data Industry Development Plan (2016–2020) emphasizes that 'data is a fundamental

[1] Bonina, C. M., New Business Model and the Value of Open Data, http://pdfs.semanticscholar.org/ceaf/d56a1172833a74246f4f3084b17d5e8b54af.pdf.

national strategic resource and a 'diamond mine' of the twenty-first century.' It can be said that there is a considerable urgency to treat government data as a 'resource' or 'asset' in theory and practice and to emphasize the benefits of preserving and increasing the value of data in the process of operation and maintenance management. This chapter proposes that, in this sense, the so-called asset-based management of government data refers to the realization of the economic and social value of government data resources in their operation and maintenance, with the operation of market mechanisms as the background condition.

Analysis of the Asset Attributes of Government Data

(1) **The asset character of government data**. Data is a carrier; its value is reflected in the information it carries or conveys. The relationship between 'data' and 'information' is understood differently in academic circles. Usually, data is regarded as the basic unit of information, and 'information is greater than data.'[2] However, from the perspective of highlighting the physical form of data, i.e., emphasizing that data is information presented in physical forms such as digital, text, image and audio, then data and information are 'form' and 'content.' Data is the form of information, and information is the content of data; data that does not convey information is like form without content. In the understanding of this chapter, therefore, government data, or government data resources, is the vehicles that convey information about the functioning of government and public power. The importance of information cannot be overstated, as the historical giant Napoleon Bonaparte said in the early nineteenth century: 90% of warfare depends on information; the contemporary management scientist Michael Porter similarly emphasized that the information revolution is sweeping across economies. He thus pointed out that the ability to access and process information is the most important competitive edge.

[2] Ye, Jiyuan, Chen, Meng, Xie Huan et al., An Exploration of the Logical Relationship Between Data and Information—And the DIKW Concept Chain Model, *Chinese Journal of Library Science*, 2017(03), 38–40.

A 2014 study supported by the UK's Engineering and Physical Sciences Research Council (EPSRC) found the lowest consensus between Napoleon and Porter: data and information can help people win wars, whether in a real or metaphorical sense, 'because data and information have such value and power that we call them 'assets' (assets).'[3] Interestingly, this study does not go to great lengths to define what constitutes an 'asset'; it seems to claim that 'something of considerable value' is an asset. The chapter argues that this vague treatment is consistent with the specific definitions of 'asset' in different disciplines such as economics and accounting[4] as they all emphasize that an asset is a 'valuable resource' that, under market conditions, has the definition of 'asset' is consistent with different disciplines such as economics and accounting, as they all emphasize that an asset is a 'valuable resource' which, under market conditions, has the potential to generate some economic or social return for the holder.

Government data resources as assets have the following characteristics:

> **Public**. As government data resources are data and information generated, collected and used by the government performing its duties, they are necessarily 'public.' China's existing administrative regulations confirm the public nature of government data resources, such as Articles 5, 6 and 7 of the Regulations of the People's Republic of China on the Disclosure of Government Information and Article 42 on the disclosure of government data upon application. From the perspective of assets, the public nature of government data resources is mainly reflected in three levels: firstly, in terms of ownership of assets, government data resources are assets owned by the public and managed by the government and its functional departments on behalf of the public; secondly, in

[3] Borek, A., A. K. Parlikad, J. Webb and P. Woodall, *Total Information Risk Management: Maximizing the Value of Data and Information Assets*, Waltham: Elsevier Inc. 2014: 4.

[4] Ma, Dung, Yu, Xin, Data Assets: Conceptual Evolution and Measurement Methods, *Journal of Statistics*, 2020(02), 15–24.

terms of management of assets, the government and its functional departments only manage them on behalf of the public and therefore must implement the basic principle of 'unified open platform, with openness as the norm and non-openness as the exception.' Third, in terms of value-added assets, the opening of government data resources to society should be free of charge. However, in order to promote the use of certain types of data resources that are highly specialized and of high relevance, the government and their functional departments have the responsibility to set up diverse incentives to encourage large enterprises or social organizations to develop and reuse such data for public interest purposes. The public nature of government data resources should be translated into the design of mechanisms for their asset-based management, which are discussed later in this chapter.

Immateriality. Data is a special kind of resource. The asset value of data resources is, on the one hand, necessarily dependent on a carrier (e.g., CD-ROM, file, digital, program, database and so on). However, on the other hand, its value is not necessarily related to that carrier. More precisely, the immaterial nature of data resources emphasizes that 'the value of data lies in its information and meaning.'[5] This means that, firstly, in the legal system, the protection of the ownership of data resources should be distinguished from the protection of the ownership of data carriers; secondly, in terms of operation and management, data resources should generally be managed in such a way as to facilitate the sensational of their information value. Therefore, managing data resources must be a 'process and systematic management of the creation, acquisition, organization, storage, distribution and use of information,'[6] rather than conventional management based on the physical form of the carrier.

[5] Glazer, R., Measuring the Value of Information: The Information-intensive Organization, *IBM Systems Journal*, 32(1), 1993, 100.

[6] Detlor, B., Information Management, *International Journal of Information Management*, 30(2), 2010, 103.

The duality of value. Generally speaking, data resources and their information products are what economists call 'empirical products': consumers must try the product each time to evaluate it.[7] For example, in the case of the most common information products such as newspapers, the fact that a high value news item was published the previous week does not guarantee that the quality of the information will be rated equally high this week. Each experience of a different individual generates a unique evaluation of value. However, the value of data resources is not entirely subjective; the ease of access, the cost of exploitation, the form of presentation, the value of associated products, etc., all have a certain degree of objectivity. Therefore, government data resources are both subjective and objective in nature, 'data is an objective existence independent of the observer, but the value of its information varies from person to person.'[8] This poses certain difficulties in pricing data resources and their products.

Renewability. Forbes.com, a leading financial information portal, published an article on 2 April 2012 entitled 'Is Data the New Oil?' The author questioned the current rampant claims that big data can create vast social and economic wealth, arguing that developing value from data should be an ongoing, cumulative process, rather than treating it as if it were a physical commodity that can be automatically realized through free trade.[9] This reflexive observation deserves to be taken seriously. In fact, by emphasizing the asset character of data resources, we are not assuming that data can 'naturally' generate significant economic and social value, but that 'the operation of market mechanisms as a background condition' is always a fundamental institutional requirement that cannot be avoided. It has become common knowledge

[7] Shapiro, Karl, Hal Van Rijn, *Information Rules: A Strategic Guide to the Network Economy*, translated by Meng, Zhaoli and Niu, Luqing, Peking: Renmin University of China Press, 2013: 4–5.

[8] Mingers, J., *Realizing Systems Thinking: Knowledge and Action in Management Science*, Berlin: Springer, 2006: 17.

[9] Palmer M., Data is the New Oil, http://ana.blogs.com/maestros/2006/11/data_is_the_new.html.

that different open applications and processing methods can produce data and information products of widely varying value. This characteristic of reproducibility shows that data resources are not only a basic resource capable of generating new values but also capable of generating different values in different ways in the hands of different subjects. In this sense, the principle of 'information justice' emerges the need to effectively guarantee that government data resources are more fully accessible and utilized by the general public, so that the value of data resources can be realized in a more diverse and multidimensional way.

(2) **Asset attributes of government data resources.**

Definition of assets under accounting standards. The implementation of asset-based management of government data requires a discussion on how it can interface with the current accounting system; that is, it needs to be possible to incorporate government data management into current accounting standards. The discussion on this issue is still very much lacking, but the need for it has already arisen in practice. This chapter first clarifies the definition of an asset under current accounting standards and then examines the type of asset that government data resources can be classified as depending on their characteristics.

Paragraph 1 of Article 27 of the Government Accounting Standards—Basic Standards issued by the Ministry of Finance on 23 October 2015 (effective 1 January 2017)—stipulates that an asset, which is one of the elements of government financial accounting, 'is an economic resource that is formed from past economic operations or events of the government accounting entity (b) Assets are economic resources that are under the control of the government accounting entity and are expected to generate service potential or an inflow of economic benefits.' (The concepts of service potential and inflow of economic benefits are explained in the second and third paragraphs of Article 27, respectively.) Article 29 adds: 'An economic resource that meets the definition of an asset under Article 27 of this Standard is recognized as an asset when both of the following conditions are met: (i) it is probable that the service potential associated with the economic resource will be realized or an inflow of economic benefits to the government accounting entity is probable; and (ii) the cost or value of the economic resource can be measured reliably.'

Classified according to liquidity, assets include current assets (according to Article 28(2), 'assets expected to be consumed or realizable within one year, including monetary funds, short-term investments, receivables and prepayments, inventories and so on') and non-current assets (according to Article 28(3), 'assets other than current assets, including fixed assets, construction in progress, intangible assets, long-term investments, public infrastructure, government reserve assets, cultural relics and assets, housing and natural resources assets and so on').

According to the 'Government Accounting Standards—Intangible Assets (Exposure Draft)' published by the Ministry of Finance, intangible assets are 'identifiable non-monetary assets that do not have a physical form but can provide a right to the government accounting entity' (Article 2). The ASBEs have come into effect also state and 'An intangible asset is an identifiable non-monetary asset that is not in physical form and is owned or controlled by an enterprise. An asset meets the identifiability criteria in the definition of an intangible asset if it meets one of the following conditions: (i) It can be separated or divided from the enterprise and can be sold, transferred, licensed, leased or exchanged, either alone or together with related contracts, assets or liabilities. (ii) Arising from contractual or other legal rights, whether or not they are transferable or separable from the business or from other rights and obligations' (EITF 6—Intangible Assets, Article 3).

In addition, two conditions must be met for the recognition of an intangible asset for accounting purposes: (1) it is probable that the service potential associated with the intangible asset will be realized or that the economic benefits will flow to the government accounting entity, and (2) the cost or value of the resource can be measured reliably.

The above conceptualization makes it difficult for government data to fall into the established asset classes under accounting standards.

Government data cannot be classified as an intangible asset under accounting standards. It is difficult to find a place for government data resources under the current financial accounting accounts. In appearance alone, government data resources are similar to 'intangible assets' in the context of non-current assets due to their distinctive immaterial characteristics. Indeed, some scholars have argued that government data resources are an intangible asset. However, careful analysis reveals many problems with this assertion.

Professor Lai Maosheng of the School of Information Management at Peking University has argued that information resources are an essential component of intangible assets. His main reasons are: that some of the main characteristics of information resources are consistent with those of intangible assets, such as 'immateriality,' 'dependence and utility'; some of the characteristics are not contradictory to those of intangible assets, such as 'cost structure specificity' and 'experiential'; some of the characteristics are partially compatible with those of intangible assets, such as 'ease of sharing'; information resources also meet the characteristics of 'uncertainty' and 'sociality' of intangible assets.[10] However, it is difficult to prove that the two types of items, which have a number of characteristics in common, belong to the same category through the method of feature comparison. For example, even Professor Lai himself argues that information resources are very different from traditional intangible assets regarding their 'transferability' and 'reproducibility.' Moreover, Professor Lai and I agree that reproducibility is one of the fundamental characteristics of data resources.

This chapter argues that government data resources cannot be recognized as intangible assets for the following principal reasons (see Table 11.1 for a detailed comparison of the two concepts):

> First, the value of government data resources lies in their 'service potential' in terms of information and meaning rather than in providing the government with certain rights. Studies have listed that data resources can provide quantitative and qualitative assessments of public administration, enterprise management or business processes, functions and the state of development; they can avoid ambiguity between different information users; they can serve as benchmarks for products and services; they can improve the efficiency of communication and decision-making within organizations;

[10] Lai, Mao-Sheng, Gao Xin-Min, Zhao Xiao-Fan et al., Information Resources: New Factors of Production, Intangible Assets and Social Wealth, *China Informatization*, 2008(10):16–17.

they can support collaboration between public functions, product data management (PDM), enterprise ERP, CRM, MRP and other functional applications.[11] In summary, government data resources have the 'service potential' to help the government accounting entity provide a public good or perform a public function. They therefore can be recognized as an asset for accounting purposes, but not through the provision of a right to the government to realize this service potential, and therefore do not meet the requirements of GASB—Intangible Assets (Exposure Draft), section "Analysis of the Asset Attributes of Government Data", which defines intangible assets.

Secondly, the public nature of government data resources (the principle of managing or protecting government data is 'owned by all and managed by the government,' which will be discussed later in this chapter) determines that it cannot simply be traded directly to realize its value (open government data must be free and open as the norm, with restricted or no openness as the exception). The way to realize the value of government data resources is through innovative open mechanisms that facilitate the use of data information by public or private decision-makers to improve the quality of their decisions. Therefore, the economic benefits generated with the cooperation of government data resources are more of a public or social benefit, distributed among different groups of enterprises or society, and cannot be identified as flowing to the government. For example, a patent right held by the government can either enable the government to increase a sum of fiscal revenue through intellectual property transactions, or attract social capital to participate in the transformation of the results through the PPP model, creating a new product in which the government enjoys certain ownership rights. However, the main criterion for judging the value of urban transport data resources is 'whether it contributes to the quality of public or private transport-related decisions.' This means that, on the one hand, there can be no direct trade in transport data and, on the other hand, there can only be innovations in open mechanisms to facilitate the access and reuse of such data, and its products by a

[11] Borek, A., A. K. Parlikad, J. Webb and P. Woodall, *Total Information Risk Management: Maximizing the Value of Data and Information Assets*, Waltham: Elsevier Inc. 2014.

larger group of people. Therefore, we believe that assets with such characteristics cannot be classified as intangible under the current definition.

Establishing a new asset class: 'data resource assets.' This chapter proposes that a third category of asset, government data resources, could be named 'data resource assets' as a topic for innovation in accounting systems.

A. **Definition**. A data resource asset is an asset that takes some form of data and that can help a government accounting entity achieve some service potential or provide some public benefit to a government accounting entity based on the use of its information.

Table 11.1 Key differences between data resource assets and intangible assets

Data resource assets	*Intangible assets*	*Compare*
Immaterial	Non-physical	Data resource assets take the form of some kind of data and cannot exist as mere 'information' separate from a specific carrier. In contrast, intangible assets can be separated from the physical carrier, for example, in the form of proof of ownership
Public utility	Public utility and economic benefits	Data resource assets are designed to achieve and promote public utility. The economic benefits derived from that information cannot be considered government assets; intangible assets, in addition to having public utility, generate economic benefits that are counted as government assets
Shareability	Monopolistic	The very purpose of open data assets is to facilitate the use of information and the development of information products; and the ownership of intangible assets is monopolistic and exclusive
Renewability	None	New data resource assets can be generated by reusing data resource assets, whereas most intangible assets do not have this reproducibility. This difference in nature also determines that the means of collecting, managing, maintaining and using data resource assets differ from intangible assets

B. **Recognition conditions**. A data resource asset is recognized if both conditions are met: The service potential arising from asset will probably be realized, and the cost or value of the asset can be measured reliably.

As the issue of asset attributes of government data resources discussed in this section is still far from having authoritative conclusions in practice and theoretical research, and as the discussion in this section is more complex, it is necessary to conclude this section with a brief overview of several of the above ideas.

For one, according to the government accounting standards, the conditions for an intangible asset to be established include: (1) it is probable that the service potential associated with the intangible asset will be realized, or it is probable that the economic benefits will flow to the government accounting entity; and (2) the cost or value of the resource can be measured reliably. According to the 'Government Accounting Standards—Intangible Assets' (Exposure Draft) launched by China Accounting Standards, 'Intangible assets for the purpose of this Standard are identifiable non-monetary assets that do not have a physical form but can provide a government accounting entity with certain rights, generally including patent rights, Trademark rights, copyrights, land use rights, non-patent technologies and other property rights, etc. Self-created goodwill and internally generated brands and newspaper names of government accounting entities should not be recognized as intangible assets.'

The definition of an intangible asset in accounting standards can be summarized as an asset that realizes some service potential for the government by providing some rights and for which the cost can be reliably measured.

Secondly, this chapter argues, firstly, that government data resources have a particular service potential and that their cost or value can be reliably calculated. Therefore, government data resources are 'close' asset to an intangible asset. The service potential of government data resources includes, but is not limited to, the ability to quantitatively and qualitatively assess public management, public decision-making, or government-supported economic and social construction activities; the ability to help different information users, such as government departments, enterprises and institutions undertaking public management functions, to avoid ambiguity in the performance of their duties; and the ability of data resources to serve as benchmarks for the public goods and services provided by the government. Data resources can be used as benchmarks for assessing the value of public products and services provided by the government; they can improve the efficiency of communication and decision-making within government organizations; and they can support various functional applications such as collaboration among public functions, data management of public products and services, economic and social development planning, and coordination between the government and citizens.

Secondly, how government data resources realize their service potential is by providing some foundational support for public activities such as government decision-making, rather than by providing some legally recognized right to do so. For example, an extensive database developed by a government realizes its service potential by helping the government to deliver the quality of certain types of public decisions through the information that can be mined from the data resources in the database; in helping to improve the quality of public decisions, the data itself, which is the underlying supporting force, also has the potential to be updated and expanded (i.e., the renewable nature of the data resources). A government patent, on the other hand, takes the form of a legally recognized right to realize a specific service potential for the government; the patented technology itself, which is the vehicle for that right, does not assist in public decision-making, nor does it have the opportunity to be renewed in the process of realizing the service potential. Therefore, government data resources cannot be categorized as intangible assets.

Admittedly, there is room for further discussion of the above discussion. The link between big data and public policy-making is still emerging. In reality, no country or region has yet dealt with the asset attributes of data resources in accounting terms, so the discussion in this chapter is intended as a 'thought experiment.' However, as a comparative reference, we can draw some approximate inspiration from the EU law on database protection.

The creation of databases can be extremely helpful in promoting economic and cultural development and scientific and technological progress. However, the construction of a database itself needs to have the 'originality' required by traditional intellectual property rights. How, then, is the legal nature of databases to be determined and how are they to be legally protected? The Directive of the European Parliament and of the Council on the legal protection of databases introduces the concept of a 'sui generis right.' In Article 7(1), the Directive states that a substantial investment in a database by the producer is a condition for the protection of a particular right, generally irrespective of the existence of originality. Substantial investment includes investment in human, financial, technical and time resources.[12]

The progressive nature of this EU law is precise: the value of copyright lies in the 'originality' required by the copyright itself, such as the intellectuality of a work or even its style of language expression; the value of a database lies in its 'usefulness,' i.e., in its contribution to other economic, cultural or scientific work. The technical content of the database design and the sophistication of the data processing procedures are independent of the essence of the database function.

We, therefore, believe that the EU's concept of 'special rights,' when viewed in the context of the asset attribution of government data resources, can lead to the inspiration that data resources should be singled out in terms of asset attribution in order to highlight their distinctive function of realizing their service potential.

Designing Principles of Ownership of Government Data Resources

(1) **Disputes caused by unclear data ownership**. The process of collecting, aggregating, opening up and utilizing big data involves multi-step operations of multiple subjects, and unclear data ownership could be more conducive to the development of the industry. In the process of data collection and open use, the conflict between the data ownership claimed by different subjects and the privacy rights of citizens has already emerged.

At the beginning of 2016, an article published in the 'Kedi Community,' 'Investigation Report on the Illegal Collection and Use of Citizens' Information on Ali's Platform,' was widely circulated online. The article criticized 'Ma's so-called big data, the essence of which is the collection, storage and use of citizens' personal information.' The article argues that although the Ali platform (including Taobao, Tmall, Alibaba and so on) includes

12 Derclaye, E., *The Legal Protection of Databases: A Comparative Analysis*, Edward Elgar Pub, 2008: 43.

procedures that require users to read and agree to the terms of an 'authorization to use' agreement before registering, and given that users register to use the Ali platform for a specific purpose, the latter can only collect and use the information within the scope of the user's purpose. The Ali platform, however, shares a large amount of data collected from users, including basic personal information and service process information, to a wide range of third parties, which not only violates the privacy rights of citizens, but also allegedly leaks state secrets. While the article's points above are open to discussion, the questions it raises are authentic, exceptionally 'Should Ali's sharing of users' personal information be constrained by users registering and using the Ali platform for specific purposes?' This question is well worth exploring in depth.

There have been cases where users have challenged the legality of online platforms collecting their records and information. In 2013, a Nanjing citizen sued Baidu Inc., because 'Baidu's search engine had recorded and tracked his own searches, violating his personal privacy.' According to a report in the Yangtze Evening News on 1 July 2015, in the second trial of the case, the court held that 'the terminal from which Baidu collected and pushed the information was the browser, and there was no targeted identification of the Internet users using the browser' and that 'Baidu did not and There was no need to link the search terms to Ms. Zhu's personal identification information,' and thus did not constitute an invasion of privacy. However, the court also recognized that 'the records of search terms formed by Internet users through the use of search engines' have privacy attributes. Therefore, it could be argued that the collection of personal information by online platforms still posed a possible risk of invasion of privacy. Although the case ended in the individual user's favor, it still did not address the fundamental issue of data ownership in terms of jurisprudence.

The challenges encountered by companies such as Ali and Baidu are, to a certain extent, the same issues that party and government departments and enterprises and institutions with public administration functions must consider when collecting and processing government data resources. There is one example of a genuine dispute: the Shanghai Stock Exchange Information Network Co., Ltd. had developed an after-hours trading statistics product, Topview, which provided data consisting of accurate trading data to seven information vendors, including Compass, who analyzed and processed the data and provided it to investors who paid a service fee. The latter analyzed and processed the data and made it available to investors who paid for the service. The CSRC ordered the SSE to shut down the database in January 2009 after receiving reports that Winful Data was providing information to the exchange for a fee that the exchange was legally required to disclose. Even so, the dispute over the data ownership, in this case, has yet to be definitively concluded. The data resources provided by Winfull Data were based on investor quotes, which were sorted, arranged and processed by the stock exchange, and then added to the extensive professional work of the information provider. From this point of view, it is debatable whether the ownership of the data belongs to the investor or the exchange, and whether the relevant data information belongs to the 'live quotations of securities transactions,' which is a matter subject to statutory disclosure by the exchange under Article 113(1) of the Securities Law.[13]

China still needs to clarify the issue of data ownership at the legal level. In 2015, the Office of the Leading Group for Network Security and Informatization of the Central Committee of the Communist Party of China (CCP) issued the Opinions on Strengthening Network Security Management of Cloud Computing Services in Party and Government Departments (Zhongwangban Fafawen [2015] No. 14); one of the basic requirements put forward is that 'data and equipment provided to service providers by party and government departments and other resources, as well as data and documents collected, generated and stored during the operation of party and government business systems on the cloud computing platform belong to the party and government departments. The service provider shall guarantee the party and government departments' access to, use of, and disposal of these resources; shall not access, modify, disclose, use, transfer, or destroy the party and government departments' data without the authorization of the party and government departments; and shall do a good job of transferring and clearing the data, documents, and other resources

[13] Wang, Yang, An Analysis of the Ownership of Securities Trading Information and its Legal Protection Model—From the Winful Data Case, *Special Economic Zone*, 2013(04), 157–158.

as required when the service contract is terminated.' This text is not controversial if it is only understood as an essential requirement to strengthen the security management responsibility of party and government departments and that constraints can be imposed on service providers through contracts by this requirement; however, if it is taken as a rule to resolve data ownership issues, it raises many questions. For example, a researcher at the China Information Security Research Institute argues that this essential requirement of the CNNIC Document 14 can hardly cope with the complexity of the practical situation: 'For the large amount of operational data of cloud computing platforms, such as system logs, should they be regarded as the data of party and government departments?'.[14]

Disputes over data ownership are not only limited to the above cases and policy disputes, but also the well-known disputes between CaiBird and Shunfeng, Sina and Stranger, Ali and Meituan and so on, all of which involve the ownership of rights to big data and the scope and limits of the use of big data in competition law. In short, there is still much room for discussion on the nature of the ownership of government data, and it cannot simply be interpreted as 'whoever owns the data (or whoever has the right to open it up).'

(2) **The nature of common property rights in data resources**. There are two reasons why data ownership is challenging to clarify: first, the tradition of the political economy holds that the source of rights is the input of labor, known as the labor theory of property rights. However, no matter which theory of labor property rights, or the Marxist labor value theory criticizes labor property rights, it is clear that the basic framework is the dichotomy of 'subject-object'; that is, when judging the attribution of rights, it is necessary to determine 'which subject, to what object, has invested labor.' The question of 'which subject, to what object, has invested the labor.' In the process of forming big data, the subjects involved in the process of data collection and processing and open use are multiple. Their activities often intersect with each other, and the information products also present an ecosystem pattern

[14] Zuo, Xiaodong, Exploring Some Technical Issues in the Network Security Management of Cloud Computing Services, GlobeNewswire: http://world.huanqiu.com/hot/2015-07/7135068.html.

> with complementary functions and mutual support, so it is not easy to clearly distinguish 'which subject has rights over which object'; secondly, according to the traditional civil law Secondly, according to the principle of 'pre-possession of unowned objects' in the traditional civil law, the ownership of the movable property can be acquired according to the fact that the movable property is possessed before others. However, the open use of big data is a process of constantly adding materials and resources. The data resources or information products that are eventually formed are also the result of multi-step operation. It is challenging to establish a time point for the formation process of big data from 'ownerless' to 'owner.' It is extremely difficult to establish a point in time when the formation process of big data changes from 'ownerless' to 'owner,' and therefore, the principle of pre-emption in civil law cannot be applied, not to mention that the principle of pre-emption has not yet been established in the legal system of civil law in China. These two reasons dictate that, as government data is often formed by multiple subjects overlapping in multiple operational steps, granting any subject (whether private or public) an exclusive property right is inappropriate.

As mentioned above, 'data is the carrier of information' and requires first focusing on the exchange and sharing of information and its development and application when thinking about data ownership. The incentives for Big Data practitioners must be given due consideration. In this regard, this chapter devises a system of data ownership principles of 'common property rights,' to strengthen the public nature and sharing of government data resources and promoting the reproducibility of the value of government data resources:

The core tenure principle: the ownership of government data resources is shared by all people, and the process of their generation, operation and maintenance management relies primarily on public financial support. The essential spirit of this principle is reflected in the current legal system, such as Article 11 of the Law of the People's Republic of China on Surveying and Mapping, which states that 'basic surveying and mapping are public welfare undertakings'; and Article 14, which states that 'people's governments at or above the county level shall incorporate basic surveying and mapping into their national economic and development annual plans and financial budgets.'

Openness principle: Party and government departments, enterprises and institutions undertaking public administration functions are responsible for the management and (unified) open use of government data formed within the scope of their statutory duties and activities, except for those involving national security and citizens' privacy.

Principle of public benefit: The opening and use of government data resources must be aimed at promoting social benefits, enhancing the public decision-making capacity of party and government departments, supporting commercial activities with significant social value or helping citizens to exercise their legitimate political and economic rights more effectively.

The principle of non-exclusivity: no single subject, whether a party or government department or an enterprise or institution or individual, can claim exclusive rights over government data resources.

Pricing Methods for Government Data Resources

(1) **Several pricing methods are currently under discussion**. According to the report 'Exploring Key Issues in China's Data Economy Development,' compiled by the Ali Research Institute in 2015, three types of data pricing methods are currently under discussion: firstly, in rem pricing, i.e., cost pricing plus brand pre-pricing. This is the pricing approach adopted by the first big data trading platform in China, Data Hall, in its business practice; secondly, intellectual property pricing, i.e., pricing according to the ownership of use, similar to the pricing model of a license. The difference between this and the 'Data Hall' pricing approach is that the former emphasizes the subjective pre-pricing by consumers; thirdly, revenue pricing, i.e., pricing by the revenue of the data. As the marginal cost of data is zero, the value generated varies with each application and person using it. This approach is generally used to value trademarks, customer relationships, proprietary technology and know-hows.

The three pricing methods mentioned above are basically based on historical trading conditions as a reference. However, as the history of data trading is still short, historical trading conditions are not sufficient to be a single reference or constraint factor for pre-pricing. At the same time, the two methods, IPR pricing and revenue pricing, are too subjective and should be considered only at a later stage. As a general rule, adding some expectation of revenue to the cost of data collection and application processing is a more operational approach to pricing. However, the cost measurement of the data must first be discussed, and current accounting standards still provide for cost measurement based on historical cost, which, concerning paragraph 2 of GASB 30, is 'measured at the amount of cash paid on acquisition or at the fair value of the consideration paid.' Paragraph 5 of Article 30: Fair value is 'the price that would be received by a market participant if the asset were sold in an orderly transaction that occurred at the measurement date.' In comparison, the acquisition of government data resources is unique. It cannot be simply applied to the provisions of accounting standards, and the provisions on historical cost and fair value must be further refined.

(2) **A comprehensive pricing approach for government data assets**. This chapter argues that the pricing of government data assets can be based on cost measurement. Yet, a benefit expectation must be added to determine the initial price of government data resources by combining elements from three different dimensions: historical factors, objective calculations and subjective evaluation. It should be noted that, due to the limitation of space and topics, this chapter only discusses data asset pricing methods within the scope of the provisions of Chinese accounting rules, with the expectation of achieving a practical interface between data asset-based management and established accounting rules. In practice, the pricing of data assets also involves many deeper theoretical issues, such as the issue of value generation of data assets, the value-added preservation of data as a renewable resource, the property rights attributes of data and the relationship with privacy rights. The General Data Protection Regulation (GDPR) has been officially implemented in the EU since 25 May 2018. The Regulation on the Free Flow of Non-Personal Data was adopted by the European Parliament on 4 October 2018. These issues are systematically addressed in documents such as the General Data Protection Regulation (GDPR), the EU Regulation on the Free Flow of Non-personal Data adopted by the European Parliament on 4 October 2018 and the California Consumer Privacy Act of 2018 (CCPA), which will

come into force in the State of California on 1 January 2020.[15] We intend to discuss these issues in a separate article.

Let's start with the question of how to measure the 'cost' of government data. There are two main types of cost measurement: historical cost measurement and replacement cost measurement. Historical cost measurement is the measurement of the amount of cash or cash equivalents paid for an asset at the time of acquisition or the fair value of the consideration paid for the asset at the time of acquisition. Historical cost measurement is based on the assumption of currency stability. It requires assets and liabilities to be recorded at their actual transaction prices at the time of acquisition or exchange, with the recorded carrying amount generally not being adjusted over the life of the asset. This measurement method is easy to operate but has obvious problems: firstly, it is difficult to truly reflect the current financial position—for example, it can be affected by factors such as inflation or price changes; secondly, it is challenging to measure assets with a highly subjective value such as data resources—for example, changes in social conditions can lead to sudden appreciation or depreciation of certain types of data resources; and thirdly, it only reflects the fact of past transactions.

Replacement cost measurement means that an asset is measured at the amount of cash or cash equivalents paid today to purchase the same or similar assets. Replacement cost measurement avoids the effects of price changes and reflects the true current financial position. However, it is problematic because it is difficult to accurately reflect the original value of the asset. For example, data resources are highly time-sensitive and the value of the same asset can vary significantly from time to time.

In order to overcome the deficiencies mentioned above in cost measurement, national accounting standards have strongly promoted fair value measurement in recent years. By fair value measurement, we mean that assets and liabilities are measured at the price that would be received to sell the asset or paid to transfer the liability in an orderly transaction between market participants occurring at the measurement date. The advantages of this approach are that it reflects current financial conditions promptly and that it can be adapted to the needs of financial innovation,

[15] Wu, Shenguo, The Global Situation of Data Governance and China's Response Strategy, *E-Government*, 2019(01), 2–10.

reflecting the market's estimate of the net present value of future cash flows that are directly or indirectly implicit in the financial instrument and facilitating accounting and innovation in financial instruments and so on.[16] However, its disadvantages are: firstly, poor reliability, assuming a high level of market maturity and equilibrium, but in practice vulnerable to human manipulation; secondly, the cost of information is relatively high, and there are very many factors to analyze; at the same time, monitoring methods to ensure fair value measurements are reliable have not yet been developed to the extent that when the sub-prime crisis broke out in 2008, many people criticized that it was the misuse of fair value measurements that led to. The financial sector has been preoccupied with the pursuit of financial innovation rather than a true reflection of financial performance.

We propose a modified 'cost plus benefit' approach to pricing based on the EPSRC (Engineering and Physical Sciences Research Council) recommended approach for enterprise information risk management, as shown in Fig. 11.1.

The cost measures in the EPSRC approach include the cost of building and maintaining the system quality, the cost of supporting services, the

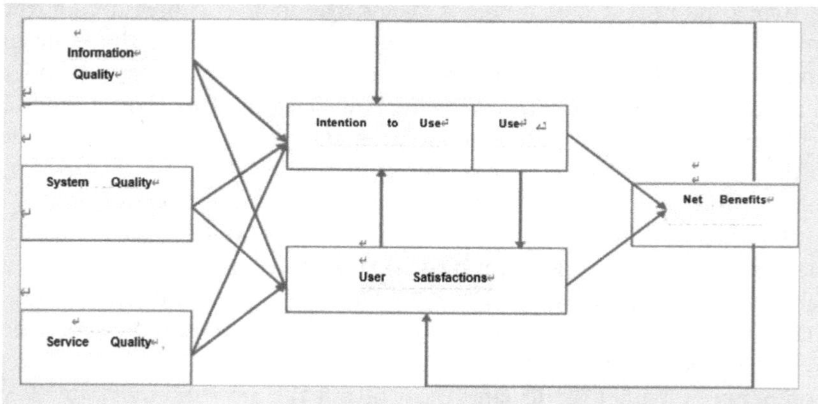

Fig. 11.1 EPSRC's recommended cost plus benefit pricing method

[16] Li Lin and Liu Yu. A comparative analysis of fair value and historical cost measurement models [J]. *Business Economics*, 2008(01), 25–26.

cost of information management and maintenance, and the expectation of benefits, including consideration of subjective user value (User Intention) and net benefits (Net Benefits). User Intention and Net Benefits are a combination of objective, subjective and historical dimensions of the product. This approach reminds us to consider the quality, scarcity and purpose of use of government data resources: firstly, because ensuring the quality of government data resources is a significant consideration in terms of input costs, it is essential to measure costs in a way that adequately reflects historical inputs and ongoing costs for maintenance and operation. Finally, as managing government data resources is to facilitate their open development, it is not appropriate to give much thought to revenue expectations.

The resulting approach to pricing government data assets can be expressed in the following non-model approach:

One, using historical cost to measure the cost of data acquisition (taking into account investment costs, tooling costs and so on.); in general, historical cost is used, and where the continuity of the amount of the asset can be assured, replacement cost may be used to measure the cost invested in securing the operation and maintenance of the data resource.

Secondly, nominal amounts are used for those measurement attributes that cannot be measured using the accounting standard categories.

Further refining, for the four main categories of government data resources, the assessment pricing methodology is categorized as follows:

(1) For government business data, the cost pricing method is adopted; that is, essential factors such as investment in the construction of hardware equipment and maintenance costs are considered.
(2) For social and public sentiment data, pricing is based on cost + moderate market per pricing. The cost consideration mainly involves the labor of non-salaried staff in data collection, the corresponding hardware and equipment construction input, maintenance costs and so on.
(3) A 'cost+moderate market' advance pricing approach is adopted for basic environmental data. At the same time, due to the significant social value embedded in environmental data, a dynamic pricing mechanism that modestly fine-tunes its market advance price according to market demand may be considered.

(4) For decentralized data, a cost pricing approach is adopted; that is, pricing is based on the cost of acquisition + the cost of maintenance and development inputs.

Conclusion: Asset-Based Management of Government Data Under the 'Collaboration and Sharing' Mechanism

Asset-based management of government data is a management method based on a complete understanding of the economic and social value of big data in government affairs. It aims to form a pattern of collaborative development and use of big data in government affairs by the government, enterprises and society, to achieve collaborative sharing and cooperative development among multiple subjects, and to work together to accelerate building a strong data nation.

In practice, many local governments in China have long explored the collaborative development and use of government affairs data, laying the foundation for theoretical discussions. For example, as early as February 2009, Peking's Leading Group of Informatization Work issued the Opinions on Strengthening the Management of Government Affairs Information Resources, which required all departments to 'implement the registration and management of government affairs information resources' and 'gradually realize the asset-based management of government affairs information resources,' initiating a policy attempt to asset-based management of government affairs data. The policy attempt of asset-based management of government affairs data was launched. At the end of 2013, Peking officially launched the government data resource network as an essential exploration of 'big data' for the benefit of the public, providing data support for the social development and utilization of government information resources. It provides data support. In 2015, the Shanghai Open Data Apps (SODA) competition was jointly organized by the Shanghai Municipal Commission of Economy and Information Technology and the Shanghai Municipal Commission of Transport, proposing that 'the aim is to explore the value of urban data and take open data as a means to explore the value of urban data. The competition aims to 'explore the value of urban data, take open data as the entry point, gather public wisdom to make suggestions for the construction of urban transportation system, and provide a competitive platform for the innovation of data products.' In April 2016, the Guangzhou Municipal Government issued the Rules for the Implementation of the Guangzhou Municipal Government Information Sharing Management Regulations, which stipulates that administrative organs may use government information obtained through sharing in accordance with the needs of their statutory functions and in accordance with statutory procedures, and in April 2020, it promulgated the Measures for Guangzhou to Accelerate the Building of an Innovative and Leading City in the Digital Economy, which proposes to 'support enterprises.' In April 2020, the Guangzhou Municipal Government promulgated the 'Measures to Accelerate the Building of an Innovative and Leading City in the Digital Economy,' which proposed to 'support enterprises, industry associations, research institutions and other social organizations to carry out data integration and applications using technologies such as 5G, big data, artificial intelligence and blockchain.'

However, in terms of asset-based management policies and systems, there are still many areas that need to be improved. On the one hand, the current government data resources are still mainly scattered under the management of various competent departments, enterprises and institutions, and social subjects, among which the development and utilization of government business data are mainly undertaken by government departments themselves or their institutions, providing services to the society without compensation or for a fee, while the use of decentralized public data for public decision-making purposes is still an uncommon exception; on the other hand, as far as the collaborative development and utilization of government business data, in the fields of traffic information, demographic information and business registration information, there have been some attempts to socialize the use of government information resources. However, due to the lack of relevant policies, incentives and restraint mechanisms, the work is still in its initial stage. Therefore, it is necessary to further promote the opening, utilization and sharing of government data by improving policies and systems for the asset-based management of government data resource and to drive the opening and utilization of various types of social data, to lay a solid foundation for the practical implementation of China's big data strategy. This chapter proposes a new asset concept of creating 'data resource assets,' the principle of data ownership of shared property rights and a comprehensive data pricing approach to establish the basic theory of asset-based management of government data.

Based on the discussion in this chapter, the direction of improving the policy and system of asset-based management of government data resources should establish two basic principles: firstly, the basic principle of data management of 'openness as the norm and non-openness as the exception' should be established; secondly, the principle of 'openness according to law and collaborative development' should be established. Secondly, the principle of sharing data resources should be established. The implementation of these two fundamental principles in the asset-based ownership rules and pricing methods for government data will also require the implementation of several institutional innovations:

(1) Provide that, in general, ownership of government data resources is shared by the public. The government is open to the general public by the purpose of public interest. Data involving national public security and confidential matters, as well as factors such as citizens' privacy and enterprises' commercial secrets, is strictly managed by relevant national laws and regulations.
(2) Study the development of a unified 'government data resource asset register,' with the data holding unit as the responsible party for management and accounting by government accounting standards. Every step of the operation and maintenance management of government data resources assets will be strictly audited and inspected to ensure the value of the assets.
(3) It is stipulated that, in general, government data is open to the public through a unified open platform. At the same time, it is stipulated that each department with management responsibility may set corresponding conditions for cooperation and development according to the degree of specialization of the data and open it up for specific public or enterprises. Products developed based on such data can be priced by the government and the developer (scientific research institutions, enterprises or other service providers) only based on the development costs invested and a limited advance price.
(4) For government data that an enterprise or other social group considers to have greater social and economic benefits but which the government does not open up, it is provided that the enterprise or other social group has the right to request the government to open up the data by the Regulations on Disclosure of Government Information and other laws and regulations by applying to a higher

government authority or the court. The enterprise or other social group accordingly pays the price determined based on the market benefits of the data (in the case of exploitation and reuse for profit) or the costs of data production (in the case of exploitation and reuse for non-profit purposes).

(5) For data resources collected and developed by enterprises or social organizations in the course of their daily operations and which are not directly applicable to the government in the performance of its public administration functions (only of a supplementary nature), government departments may legally purchase the use of such data through government procurement if they consider it necessary to do so.

(6) For data resources collected and developed by enterprises or social organizations in the course of their daily operations but which have a direct or primary role in the performance of the government's public administration functions, it is provided that government departments may purchase the use of such data through government procurement by way of negotiation. If the parties fail to negotiate a deal, the government may legally apply to the court to obtain the right to use such data by paying the price based on the cost of developing the data plus a limited advance price.

(7) Organize review panels comprising government officials from relevant departments, business people, professional accountants, scientific researchers and representatives of the general public to evaluate a sample of government data resources asset-based management on a regular and irregular basis, with a view to identifying problems and drawing lessons promptly and promoting continuous improvement of the relevant systems.

CHAPTER 12

Public Choice in Artificial Intelligence Governance Policy: Toward 'Shared Freedom'

In July 2017, the State Council issued the Next Generation Artificial Intelligence Development Plan, which not only made strategic plans for the development of AI but also established a 'three-step' policy goal to build China into the world's leading AI innovation center by 2030.[1] It is worth noting that this plan is not just about technology or industry development but also about social construction, institutional restructuring, global governance and other aspects. This is so because AI technology itself is generic and fundamental. In other words, to facilitate the rise of the AI era, we are not faced with achieving a disruptive technological breakthrough in a particular area of expertise or industry, but rather vigorously promote the comprehensive changes arising from technological development.

For this reason, the challenges faced in the development of AI are not limited to the technical or industrial sphere, but are more often reflected in public policy choices in the economic, social and political spheres. Firstly, can the governance structure of public affairs, which is generally based on a hierarchical system, adapt to the uncertainty and

[1] Notice of the State Council on the Issuance of the Development Plan for a New Generation of Artificial Intelligence, http://www.gov.cn/zhengce/content/2017-07/20/content_5211996.htm.

Y. Jiang and K. Jia, *Transformation of Platform Governance in China*, IPP Studies in the Frontiers of China's Public Policy,
https://doi.org/10.1007/978-981-99-6456-7_12

unpredictability that the development and application of technology will generate on a large scale? Furthermore, can the long-standing regulatory system-based 'command-control' mechanism around human behavior be adapted to a data- and algorithm-driven application environment? Finally, how can new governance systems and tools be constructed to address the new economic, social and political issues that will emerge due to AI development?

Addressing the above challenges only depends partially on technological development or commercial innovation per se, but more on our public policy choices. Based on analyzing the development logic of artificial intelligence and the risks and challenges caused by it, this chapter attempts to analyze the path of public policy choice in the era of artificial intelligence and discuss the possible path of future reform. Thus, three questions of artificial intelligence governance are constituted. We advocate a change in AI and platform governance toward 'shared freedom' by answering these three basic questions. In other words, AI itself has become a governance object. Its development and application pose a governance challenge, and how to make public policy choices on this basis is the direction of future governance changes.

The paper is divided into four parts: The first part examines the concept and characteristics of artificial intelligence and explains the logic of its development. As a disruptive technological innovation, its technological threshold poses a challenge to policy-makers, and the clarification of the essence of AI is, therefore, a prerequisite for formulating relevant public policies. The third part of this chapter provides a comparative overview of these challenges and points out the significance of their progress. It is important to note that although the policy objectives of each country attempt to pursue a two-dimensional balance between AI development and regulation, the current public policy choices are not comprehensive due to the lack of a complete understanding of the connotation of AI and its development logic; the fourth part of this chapter will propose new governance ideas and other possible paths for public policy choices in order to promote in-depth public policy issues related to AI governance. The fourth part of this chapter will propose new ideas on governance and other possible paths for public policy choices to facilitate in-depth discussions on public policy issues related to AI governance.

The Concept of Artificial Intelligence and Technological Development: Algorithms and Data

With the rapid development of artificial intelligence technology, especially in recent years in chess playing, autonomous driving and face recognition, there is a heated debate about the social changes that artificial intelligence may trigger. On the one hand, many people, including Stephen Hawking, Musk, Bill Gates and Hilarie, have called for more regulation to prevent 'artificial intelligence from becoming the end of human civilization'[2]; on the other hand, politicians and academics, including Barack Obama, have argued that regulation should be relaxed to unleash the full potential of AI technology for the benefit of society.[3] While the uncertainty of the future is one of the major causes of the current debate, the differences in understanding of the concept of AI and the lack of clarity of its development logic may also be a serious source of disagreement. For this reason, clarifying the conceptual content and development logic of AI is necessary to respond to the debate and as a prerequisite for further public policy proposals.

There has yet to be a general consensus on the definition of the concept of 'artificial intelligence' as far as the relevant research field is concerned. Alan Turing, a pioneer in the field of computing, argued in his article 'Computers and Intelligence' that it is not the ability of a machine to imitate human thought processes that are important but rather the ability of a machine to repeat the outwardly expressive behavior of human thought.[4] From this understanding, the famous 'Turing Test' was developed. However, as John McCarthy, a professor of computer science at Stanford University, has pointed out, the Turing Test is only one part of the concept of 'artificial intelligence,' and machines that do not mimic

[2] Stephen Hawking says A.I. could be 'worst event in the history of our civilization', https://www.cnbc.com/2017/11/06/stephen-hawking-ai-could-be-worst-event-in-civilization.html. Elon Musk, Artificial Intelligence is Our Biggest Existential Threat, https://www.theguardian.com/technology/2014/oct/27/elon-musk-artificial-intelligence-ai-biggest-existential-threat. Microsoft's Bill Gates Insists AI is A Threat, http://www.bbc.com/news/31047780. Harari, Yuval Noah, *A Brief History of Humankind*, translated by Lin, Junhong, Peking: CITIC Press, 2014.

[3] The President in Conversation With MIT's Joi Ito and WIRED's Scott Dadich, https://www.wired.com/2016/10/president-obama-mit-joi-ito-interview/.

[4] Turing, A. M., Computing Machinery and Intelligence, *Mind*, 1950, 59(236).

humans but can also perform relevant behavior should also be considered 'intelligent.'[5] John McCarthy originated the modern concept of artificial intelligence. In his view, 'intelligence' is about the behavioral 'mechanisms' that accomplish specific goals, and machines can either imitate humans to achieve behavioral mechanisms, or they are free to use any method to create them.[6] This leads us to another fundamental concept in artificial intelligence—'machine learning.'

Artificial intelligence research aims to enable machines to achieve human-level intelligence, the most important of which is the ability to learn.[7] Therefore, although 'machine learning' is a sub-domain of 'artificial intelligence,' we often equate the two concepts.[8] In terms of implementation, machine learning refers to using certain algorithms to guide a computer to derive an appropriate model using known data and to use this model to make judgments about new situations to complete behavioral mechanisms. Here, it is essential to highlight the difference between machine learning algorithms and traditional algorithms. An algorithm is a set of instructions telling a computer what to do. Traditional algorithms often specify in detail what the machine will do under given conditions; machine learning algorithms 'learn' from existing data, enabling the machine to make judgments in new situations that differ from the historical data. In a traditional algorithm, for example, the programmer carefully specifies the process for implementing each action in a given environment; in a machine learning algorithm, the programmer makes the computer analyze and simulate human walking actions so that it can walk even in a completely unfamiliar environment.

This allows us to set a 'working definition' of 'artificial intelligence' to facilitate further discussion: artificial intelligence is a computing system based on modern algorithms, supported by historical data, that can perceive, reason, learn, make decisions and other thinking activities and

[5] McCarthy, J., What is Artificial Intelligence, http://www-formal.stanford.edu/jmc/whatisai/whatisai.html.

[6] McCarthy, J., What is Artificial Intelligence, http://www-formal.stanford.edu/jmc/whatisai/whatisai.html.

[7] McCarthy, J., What is Artificial Intelligence, http://www-formal.stanford.edu/jmc/whatisai/whatisai.html.

[8] Domingoz, Pedro, *The Ultimate Algorithm: How Machine Learning and Artificial Intelligence Are Reshaping the World*, Translated by Huang Fangping, Peking: CITIC Press, 2016.

It is a computational system that is capable of perceiving, reasoning, learning, making decisions and accomplishing behavior by specific goals. Although it may still be imperfect, this concept highlights the two cornerstones of the development and application of AI technology—algorithms and data—and facilitates the discussion on the governance of AI.

Firstly, algorithms are rules that establish what the machine is trying to achieve and indicate the paths and methods to achieve it. In the current history of AI technology, algorithms can be divided into five main categories: the symbolic school, the linkage school, the evolutionary school, the analogical school and the Bayesian school.[9] Each school has followed a different logic and implemented the artificial intelligence process (i.e., 'machine learning') with different ideas. The 'symbolic school,' for example, reduces all information processing to manipulating symbols so that the learning process is reduced (abstracted) to a process of rule induction based on data and assumptions. Based on data (i.e., historical facts) and existing knowledge (i.e., pre-determined conditions), the semiotic school trains machines to learn and thus to make decisions in new contexts through a process of 'hypothesis formulation—data validation—further formulation of new hypotheses—induction of new rules.'

From the description of the 'symbolic school,' it is clear that the key to the success of a machine learning model is not only the algorithm but also the data. Lack of data and poor preconditioning will directly affect the output of machine learning (in the case of the symbolic school, the generalization of decision rules). The most obvious example of this problem is Russell's 'inductivist turkey' problem: after 10 days of observation (with an incomplete data set), turkey reaches a conclusion (representing an unreasonable pre-condition; more than ten confirmatory data means that the rule is accepted) and the owner feeds it every morning at 9am; but then it is breakfast on Christmas Eve, and instead of feeding it, the owner slaughters it. The owner did not feed it but slaughtered it.

Despite their different philosophies, all algorithm types focus on 'algorithms' and 'data' as the key to model success. If one thinks outside of specific schools of thought, each machine learning algorithm can be

[9] Domingoz, Pedro, *The Ultimate Algorithm: How Machine Learning and Artificial Intelligence Are Reshaping the World*, translated by Huang Fangping, Peking: CITIC Press, 2016.

summarized in three parts: 'representation, evaluation, and optimization.'[10] While machines can continually optimize themselves to improve their learning capabilities and can, in principle, learn anything, the methods and principles of evaluation (algorithms) and the data used to evaluate them (data) are all human decisions—and this is where the governance of AI comes into play. Algorithms and data are not only the cornerstones of the logic of AI development, but they are also the object and key to governance.

In conclusion, the debate around 'whether AI will replace humans' is not very meaningful. However, it is more important to answer the questions of 'what to govern' and 'how to govern' after clarifying the meaning of AI and understanding its development logic. The question of 'what' and 'how' should be answered—in this regard, clarifying that the objects of governance are algorithms and data is undoubtedly an important step. However, the next important question remains: What are the governance challenges posed by the rise of the artificial intelligence era? Is the current institutional design able to respond effectively to them? However, if the answer is no, how can we reconfigure our governance systems to meet the rise of the AI era? The remainder of this chapter will elaborate on this further.

Governance Challenges on the Rise in the Age of Artificial Intelligence

Unlike other disruptive technologies, the development of AI is not confined to a specific industry but is a general-purpose technology that can support the transformation of all industries. Because of this, it has a wide range of social spillover effects and will bring about profound changes in various political, economic and social fields and will also lead to governance challenges. Specifically, the challenges are manifested in three main areas.

First, the rigidity of the governance structure, i.e., the traditional hierarchical governance structure, may need help to cope with the openness and uncertainty created by the rapid development of AI. AI needs to be regulated because it can be a source of public danger, such as when

[10] Domingoz, Pedro, *The Ultimate Algorithm: How Machine Learning and Artificial Intelligence Are Reshaping the World*, translated by Huang Fangping, Peking: CITIC Press, 2016.

autonomous driving technology becomes widespread, which can lead to large-scale continuity of harm if something goes wrong. However, unlike the twentieth-century sources of public danger such as aircraft, large dams and atomic nuclear technology, the development of AI is so open-ended that any programmer or company can develop and apply AI programs without any barriers. On the one hand, this is due to the advent of the Internet era, which has allowed the threshold for code-based production to be significantly lowered[11]; on the other hand, it is also necessary for the law of development of AI itself. As mentioned above, the only way to get better machine learning results is to have large-scale data input, so it has become an inevitable choice for most AI companies to make the AI platform (i.e., algorithm) open source so that more people can use it in different scenarios and thus absorb more and more complete data to improve the algorithm itself. At the same time, the openness of the AI production model inevitably brings uncertainty. In the absence of practical constraints or guidance, the development of AI is likely to go astray. In the face of this new situation, it is clear that the traditional, section-based governance structure needs help in responding effectively. On the one hand, the government's attempts at ex-ante regulation with full coverage have become impossible, and the open AI production network makes it almost impossible for regulators to find targets for regulation; on the other hand, the top-down authority structure can neither be passed on to producers nor can the problem of information asymmetry be exacerbated, which may lead to regulatory actions going the other way. Adjusting the governance structure and logic, and developing a model of AI production adapted to the characteristics of openness and uncertainty, is one of the current governance challenges faced.

Furthermore, the lagging nature of the governance approach, i.e., the long-established system of legal regulation based on the causality of human behavior, may be challenging to apply to an application environment dominated by algorithms and data. The value of artificial intelligence does not lie in imitating human behavior, but in its ability to learn and make decisions autonomously, it is for this reason that AI technology cannot be understood simply as an expression of the will of its creator (i.e., a human being). The programmer gives the rules for learning, but the algorithm itself, trained on large-scale data, makes the

[11] Benkler, Y., *The Wealth of Networks: How Social Production Transforms Markets and Freedom*, New Haven: Yale University Press, 2006.

decisions, and this result is not directly causally linked to the will of the programmer. This feature makes it possible for AlphaGo to beat consecutive Go champions, while its designer is not a top Go master. In this sense, we come back to Foucault's notion of the 'subjectivity of technology.' For him, 'technology is not merely a tool, or a means to an end; rather, it is a political actor, where the means and the end are inseparable.'[12] In this regard, the long-standing legal system of regulating human behavior by exploring the causal relationship between action and consequence may encounter a dilemma: it is certainly not convincing to blame the designer of an AI for the infringements it causes, but how can we hold a machine accountable if we want to blame the AI itself? Thus, the second governance challenge is dealing with the allocation of public responsibility brought about by algorithms and data as the core of the technical subject.

Finally, the narrow scope of governance, i.e., the need to build new governance systems and develop new governance tools for new social issues arising from the impact of AI development. The challenges of governance arising from the development of artificial intelligence are not only the incompatibility of existing systems but also the gaps in governance for new issues. In particular, this includes the following issues: whether algorithms can enjoy constitutional protection of freedom of expression, how to define the ownership of data, how to mitigate the inequalities that AI may exacerbate and how to balance AI development with unemployment. Before the era of AI, these issues did not exist or were not prominent; however, with the rapid development of AI and its widespread application, their importance has become increasingly apparent. Regarding technological possibilities, the widespread use of artificial intelligence and robots to replace manual labor is already an undeniable fact. Whether it is a journalist, a stock market analyst or even a legal practitioner, all of them could be replaced by machines. In a world of 'Full Automation,' one of the most pressing governance challenges is to reconceptualize the relationship between labor and welfare protection and to restructure labor and welfare protection systems.[13]

[12] Foucoult, M., *Discipline and Punish*, translated by A. Sheridan, Paris, FR, Gallimard, 1975.

[13] Srnicek, N., & Williams, A. The Future Isn't Working, *Juncture*, 2015, 22(3).

Together, these three aspects constitute the governance challenges arising from the rise of the AI era. In the face of these challenges, countries have made corresponding public policy choices. The third part of this chapter will provide a comparative analysis of the governance policies of various countries on AI. Based on this, Part IV will present the policy recommendations of this chapter.

An Overview of National AI Governance Policies and Regulatory Paths

The rise of the artificial intelligence era as a widespread phenomenon has led to governance challenges common to all countries, and public policies have been introduced to facilitate and regulate the rapid development of AI.

In 2016, the US issued two national policy frameworks, the National AI Research and Development Strategic Plan and Preparing for the Future of AI, with the former focusing on the purpose, vision and priorities of the US AI strategy from a technical perspective. The latter exploring the role of government in promoting innovation and ensuring public safety from a more governance perspective. Regarding specific regulatory policies, Preparing for the Future of Artificial Intelligence proposes a general response, emphasizing the principles of risk assessment and cost–benefit considerations in deciding whether to impose regulatory burdens on the development and application of AI technologies.[14] Japan also introduced the Fifth (2016–2020) Science and Technology Basic Plan in 2016, which introduced the concept of 'Super Intelligent Society 5.0,' emphasizing the development and application of AI through various efforts such as promoting data standardization, building social service platforms and coordinating the development of multidisciplinary intelligent systems. The concept of 'Super-intelligent Society 5.0' was introduced, emphasizing the development and application of artificial intelligence by promoting data standardization, building social

[14] Preparing for the Future of Artificial Intelligence, https://obamawhitehouse.archives.gov/sites/default/files/whitehouse_files/microsites/ostp/NSTC/preparing_for_the_future_of_ai.pdf.

service platforms and coordinating the development of multi-disciplinary intelligent systems.[15]

Despite their different policy focus, the US and Japan share a generally inclusive and open attitude toward the development of AI and the challenges it poses. In terms of the current policy framework, the policy objectives of the US and Japan are more geared toward promoting technological innovation and maintaining their national competitive advantage; when it comes to regulating the public issues that AI may raise, their policy choices also tend to follow a 'permissionless' regulatory logic; that is, they emphasize the importance of regulation unless there is sufficient evidence of harm. When it comes to regulating the public issues that AI may raise, the policy choices also tend to follow a 'permissionless' regulatory logic, emphasizing that new technologies and business models are, by default, permissible unless there is solid case to prove their harmfulness.[16] As for the potential threat of AI developments to personal data privacy and public safety, although addressed in both countries' policy frameworks, it is different from the focus of their policies—in contrast to the UK and France, which have taken a different policy path.

In 2016, the UK government published 'Artificial Intelligence: opportunities and implications for future decision making,' which sets out and maps out the transformative impact of AI and how it can be used, focusing on the legal and ethical risks associated with its development. In the report, the UK government highlights the impact of machine learning combined with personal data on fundamental rights such as personal freedom and privacy, clarifies the concept and mechanism of accountability for AI decisions, and sets out specific policies on algorithmic transparency, algorithmic consistency and risk allocation.[17] Similar to the UK, France's AI Strategy published in 2017 continues the spirit of its 2006 Information Society Act, which also emphasizes the need to strengthen the 'co-regulation' of new technologies in order to enjoy the improved welfare of technological developments while fully protecting

[15] Xue, Liang, Japan Pushes to Realize Super-smart Society 'Society 5.0', http://www.istis.sh.cn/list/list.aspx?id=10535.

[16] Thierer, A., *Permissionless Innovation: The Continuing Case for Comprehensive Technological Freedom*, Mercatus Center at George Mason University, 2016.

[17] Artificial Intelligence: Opportunities and Implications for the Future of Decision Making, https://www.gov.uk/government/uploads/system/uploads/attachment_data/file/566075/gs-16-19-artificial-intelligence-ai-report.pdf.

individual rights and the public interest.[18] Compared to the US and Japan, public policy in the UK and France is more oriented toward a 'precautionary' policy logic, which emphasizes that new technologies or business models should only be allowed to be used if the developer proves they are not harmful.[19]

In this chapter, both 'regulation without approval' and 'prudential regulation' have merit in addressing the governance challenges posed by the rise of the AI era: the former focuses on promoting innovation. At the same time, the latter is more robust due to its focus on safety. It should be noted, however, that the shortcomings of both regulatory paths are also evident. As noted in Part II above, on the one hand, rapidly iterating technological developments and business model innovations will inevitably raise new social issues, whether it be the protection of algorithms under the right to freedom of expression or the social challenges posed by widespread unemployment, which objectively require public policy responses that go beyond a one-sided 'regulation without approval.' They objectively demand public policy responses beyond a one-sided 'regulation without approval.' More importantly, the underlying assumption of 'regulation without approval' is the effectiveness for ex-post regulation; however, in reality, as the 'instant crash' of the Dow Jones Industrial Average on 6 May 2010 revealed, even if a single electronic trading program is operating in compliance when individual systems behave in a. However, in reality, as the 'instant crash' of the Dow Jones Industrials on 6 May 2010 revealed, even if individual electronic trading processes operate in compliance, when the actions of individual systems are aggregated, they create an even greater crisis.[20] In such cases, 'ex-post regulation,' which relies on compliance judgments, is largely ineffective. On the other hand, the autonomous and subjective nature of AI makes the logic of 'prudential regulation' based on the causality of human behavior inherently flawed: how can developers prove the harmlessness of AI systems, given that humans cannot predict their likely behavior or decisions?

[18] Zhou, Yanbing, The Development and Application of Big Data Industry in France, *Study Times*, 2014-11-03.

[19] Thierer, A. D. & Watney, C. J., *Comment on the Federal Automated Vehicles Policy*, 2016.

[20] Kaplan, Jerry, *The Age of Artificial Intelligence*, translated by Li, Peng, Peking: CITIC Press, 2016.

As this chapter has repeatedly emphasized, what distinguishes AI from other revolutionary technologies is precisely the comprehensive and fundamental nature of the social impact it brings. Artificial intelligence is not a technological breakthrough in a single field or industry but a fundamental change in the functioning of society; the rise of the AI era is not an overnight achievement but a 'singularity' change based on the computer revolution, the Internet revolution and the digital revolution. Therefore, in the face of the governance challenges posed by the rise of the AI era, we should also develop a comprehensive public policy framework rather than just following the traditional governance logic, for example, by only addressing specific issues under the binary dimension of 'innovation' and 'security.' The choice between 'innovation' and 'security' for specific issues is difficult. In the fourth part of this chapter, policy recommendations are made in three areas: recognizing the subjectivity of technology, reconstructing social governance systems, and promoting global governance of AI, with the hope that this will contribute to more profound research and discussion on public policy choices in the AI era.

Public Policy Options in the Age of Artificial Intelligence: Toward 'Shared Freedom'

The New Generation Artificial Intelligence Development Plan sets out a 'three-step' goal for the development of AI in China by 2030. At each stage, the gradual establishment and improvement of AI laws and regulations, ethical norms and policy systems are essential elements. Faced with the challenges of governance in the era of AI, how to reconstruct the governance system, innovate governance mechanisms and develop governance tools is an essential challenge for policy-makers. Based on the analysis of the basic concepts and development logic of AI and the comparative analysis of existing policies in various countries, this chapter proposes the following three reform ideas to provide reference for public choices in the AI era.

First, the cornerstone of AI development is algorithms and data. Establishing and improving the governance system and governance mechanism around algorithms and data is the primary proposition for public policy choices in the AI era and is also an inevitable requirement for addressing governance challenges and giving subjectivity to algorithms and data. (1) Regarding the governance of algorithms, the core issue involved is the right to formulate algorithms and the corresponding supervision procedures. Who has the right to make algorithms as the main rules in the age of artificial intelligence, and through what process, and who monitors them and how? The importance of this question is illustrated by the long-standing public questioning of the social media Facebook: how can the public trust that special interest orientations will not taint the news content that Facebook automatically recommends to its users?[21] As more and more people rely on customized news feeds, artificial intelligence could influence presidential elections. This is also why many governance principles, including transparency and open-source requirements, should be considered in topics related to algorithmic governance. (2) In terms of data governance, as AI increasingly relies on collecting and using large-scale data, issues related to protecting personal privacy, distributing data value and data security will also become the focus of public policy. Balancing different value demands and regulating data sharing and application has likewise become another important grip on public policy choices in the era of AI.

Secondly, we should innovate the social governance system and further improve the social security system so as to alleviate, to the greatest extent possible, the uncertainties that the development of artificial intelligence may bring about. Similar to technological revolutions in history, the development of AI will also lead to the division and reconstruction of interests and how to ensure that those who bear the costs of technological revolution are compensated to the maximum extent and that all people enjoy the 'sense of gain' from technological development is not only a requirement for fair and just social development, but also a catalyst for the faster completion of technological revolution. It is also a catalyst for the faster completion of the technological revolution. In this regard, in considering public policies related to AI, we should not only focus on industrial and economic policies but also social policies, as only the latter can be improved to control the risks taken by workers or entrepreneurs and help them to judge whether to support or resist change. Regarding specific policy design, the general establishment of a basic income system to mitigate the potential wave of unemployment caused by artificial intelligence may be high on the agenda. A 'basic income' is paid to all members

[21] Rosenbach, Marcel, How Google and Facebook Can Reshape Elections, http://www.spiegel.de/international/germany/google-and-facebook-could-help-decide-2017-german-election-a-1120156.html.

of a political community (e.g., the state), without any restrictions, to meet their basic living needs. Despite doubts about 'lazybones,' some researchers have pointed out that the basic income system, conceived since the eighteenth century, is likely to promote employment in turn.[22] The Finnish government has already begun experiments with this in early 2017, and some states in the US and Switzerland have also explored it to some extent. Before the age of artificial intelligence has fully revealed its 'hideousness,' innovative social governance mechanisms and improved social security systems may be the best way to balance technological innovation and social risk.

Thirdly, we should build a global AI governance mechanism, promote the resolution of major international common problems in AI in various forms and jointly address the global challenges of the open AI production model. The development of AI is characterized by openness and uncertainty, and the lowering of production thresholds has led to transnational solid mobility in AI technology research and development. Formulating relevant standards, establishing open platforms and forming a framework for sharing and cooperation all require the construction of a corresponding global governance mechanism. On the other hand, the rapid development of cross-border data flow in terms of breadth and depth has become a direct driving force for the advancement of AI technology. However, the vast differences in data regulation systems among countries will also affect the overall arrival of the AI era while restricting the further development of cross-border data flow.[23] Therefore, innovative global governance mechanisms and the possibility of cooperation and sharing while acknowledging the differences between national systems are important considerations for public policy choices in the AI era. In terms of specific mechanism design, a multi-stakeholder model can be introduced into the construction of the AI global governance mechanism; conversely, to prevent the formation of giant monopolies, a multilateralist model that gives full play to the role of sovereign states cannot be ignored. As a fundamental technological change with far-reaching implications, the

[22] Van Parijs, P., Basic Income: A Simple and Powerful Idea for the Twenty-first Century, *Politics & Society*, 2004, 32(1).

[23] Jia, Kai, Global Governance of Cross-border Data Flows: Conflict of Power and Policy Cooperation: The Evolution of Regulatory Regimes for Cross-border Data Flows in Europe and the United States as an Example, *Journal of Shantou University* (Humanities and Social Sciences Edition), 2017(5).

experience and lessons learned from the global governance mechanism of the Internet are worth learning from the development of AI.

These three areas provide a holistic view of the public policy framework in the era of AI. In contrast to traditional policies that are limited to a two-dimensional choice between 'innovation' and 'security,' this chapter presents a more integrated view of possible paths for future public policy choices. In terms of their intrinsic links, the establishment and improvement of governance systems around algorithms and data are the starting point, which will reconstruct the rules and institutions in the era of AI; the innovation of social governance mechanisms and the improvement of social security systems are the bottom line, which will mitigate the impact and volatility brought about by AI; and the construction of global governance mechanisms becomes the institutional infrastructure on which countries can move together toward a 'community of human destiny' in the era of AI—The 'community of human destiny.'

Concluding Remarks

After more than 60 years of development, artificial intelligence has finally taken off based on many technological breakthroughs, such as the Internet, big data and machine learning. In the future, AI will also play an increasingly important role in human life. We should not be alarmed by such a scenario but should not take it lightly. Finding the right direction and taking the proper measures for the governance of AI is a policy issue that needs to be addressed now. This chapter aims to break through the long-standing 'general' concerns about AI, identify the technical logic of AI technology development and the governance challenges it poses, and propose policy options. These three fundamental questions of AI governance are the prerequisites for the reconstruction of the governance system, the innovation of governance mechanisms and the development of governance tools. Along introducing China's national-level strategic plan, the development of AI in China is bound to leap to a new level. In this context, in-depth discussion of public policy issues related to artificial intelligence governance is highly urgent and for promoting the rapid development of the artificial intelligence industry, and empowering ordinary people and realizing 'shared freedom' through this development.

References

Abbate, Janet, *Inventing the Internet*, Cambridge, Mass.: MIT Press, 1999.

Abramson, Bruce, *The Digital Phoenix: Why the Information Economy Can Be Reborn*, translated by Zhao, Pei, and Zheng, Xiaoping, Shanghai: Shanghai Far Eastern Publishing House, 2008.

Acemoglu, Daron, & Pascual Restrepo, Robots and Jobs: Evidence from US Labor Markets, https://conference.nber.org/confer/2017/PRs17/Acemoglu_Restrepo.pdf (Last accessed: 10 June 2022).

Acemoglu, Daron, and Pascual Restrepo, The Wrong Kind of AI? Artificial Intelligence and The Future of Labor Demand, NBER Working Paper Series, http://www.nber.org/papers/w25682.

Ackerman, Bruce, and James S. Fishkin, Deliberation Day, *The Journal of Political Philosophy*, 2002, 10(2).

Allen, Robert, *The British Industrial Revolution in Global Perspective*, translated by Mao, Likun, Hangzhou: Zhejiang University Press, 2012.

Andreas, Joel (An Zhou), *The Rise of the Red Engineers: Tsinghua University and the Rise of China's Technocratic Class*, translated by He, Daming, Hong Kong: Chinese University of Hong Kong Press, 2017.

Atkinson, Robert D., Who Lost Lucent?: The Decline of America's Telecom Equipment Industry, https://americanaffairsjournal.org/2020/08/who-lost-lucent-the-decline-of-americas-telecom-equipment-industry/.

Baiocchi, Gianpaolo, Participation, Activism, and Politics: The Porto Alegre, in Archon Fung and Erik Olin Wright, eds., *Deepening Democracy. Institutional Innovations in Empowered Participatory Governance*, London and New York: Verso, 2003.

Y. Jiang and K. Jia, *Transformation of Platform Governance in China*, IPP Studies in the Frontiers of China's Public Policy,
https://doi.org/10.1007/978-981-99-6456-7

Barlow, John Perry, A Declaration of the Independence of Cyberspace, https://www.eff.org/cyberspace-independence.

Benkler, Y., *The Wealth of Networks: How Social Production Transforms Markets and Freedom*, New Haven: Yale University Press, 2006.

Ben-Shahar, Omri, and Karl E. Schneider, *More Than You Wanted to Know: The Failure of Mandated Disclosure*, translated by Xiaofang Chen, Peking: Law Press, 2015.

Berberich, Matthias, and Malgorza Tastein, Blockchain Technology and the GDPR—How to Reconcile Privacy and Distributed Ledgers, *European Data Protection Law Review (EDPL)*, 2016, 2.

Birtchnell, Thomas, William Hoyle, *3D Printing for Development in the Global South: The 3D4D Challenge*, London: Palgrave Macmillan, 2014.

Boix, Carles, *Democratic Capitalism at the Crossroads: Technological Change and the Future of Politics*, Princeton and Oxford: Princeton University Press, 2019.

Bonvillian, William B., and Peter L. Singer, *Advanced Manufacturing: The New American Innovation Policies*, Cambridge: The MIT Press, 2017.

Borek, A., A. K. Parlikad, J. Webb, and P. Woodall, *Total Information Risk Management: Maximizing the Value of Data and Information Assets*, Waltham: Elsevier Inc., 2014.

Bory, Paolo, *The Internet Myth: From the Internet Imaginary to Network Ideologies*, London: University of Westminster Press, 2020.

Braudel, Fernand, *The Character of France: People and Things* (vol. 2), translated by Gu, Liang and Zhang, Zeqian, Peking: The Commercial Press, 1997.

Braverman, Harry, *Labour and Monopoly Capital: The Degradation of Work in the Twentieth Century*, New York and London: Monthly Review Press, 1974.

Brown, Ian, Interoperability as a Tool for Competition Regulation, OpenForum Academy, Nov. 2020.

Brynjolfsson, Erik, and Andrew McAfee, *The Second Machine Age: Work, Progress, and Prosperity in a Time of Brilliant Technologies*, New York: W. W. Norton &Company, Inc., 2014.

Burawoy, Michael, *The Politics of Production: Factory Regimes Under Capitalism and Socialism*, London: Verso, 1985.

Buterin, V., Visions Part I: The Value of Blockchain Technology, https://blog.ethereum.org/2015/04/13/visions-part-1-the-value-of-blockchain-technology/.

Calabresi, Guido, and A. Douglas Melamed, Property Rules, Liability Rules, and Inalienability: One View of The Cathedral, *Harvard Law Review*, 1972, 85.

Cameron, Charles M., *Veto Bargaining: Presidents and The Politics of Negative Power*, Cambridge: Cambridge University Press, 2006.

Cameron, Nigel, *Will Machines Take Your Job?*, translated by Wei Qian and Wang Litao, Peking: China Workers Press, 2020.

Chander, Anupam, How Law Made Silicon Valley, *Emory Law Journal*, 2013, 63.

Chandler, Alfred D., James W. Cortada, eds, *Information Changed America: The Forces Driving National Transformation*, translated by Wan Yan and Qiu Yanjuan, Shanghai: Shanghai Far Eastern Publishing House, 2011.

Chen, Duxiu, The Clintide Tablet, *New Youth*, vol. 5, no. 5 (15 October 1918).

Chen, Duxiu, Reply to the Crime of the New Youth, *New Youth*, vol. 6, no. 1 (15 January 1919).

Cui, Zhiyuan, The An-gang Constitution and Post-Fordism, *Reading*, no. 3, 1996.

Cui, Zhiyuan, Marx, Tocqueville and Proudhon in 1848, *21st Century* (Institute of Chinese Culturle, Chinese University of Hong Kong), 2018(6).

Cui, Zhiyuan, Greater Convergence More Likely than Decoupling, *China Daily*, October 9, 2019.

Cui, Zhiyuan, Double-Cycling, Global Value Chains and Additive Manufacturing: Preliminary Theoretical Reflections on New Development Strategies, *21st Century* (Institute of Chinese Culture, Chinese University of Hong Kong), August 2021 issue, General 186.

Cukier, Kenneth, and Viktor Mayer-Schoenberger, The Rise of Big Data: How It's Changing the Way We Think About the World, *Foreign Affairs*, vol. 92, no. 3 (May/June 2013).

D'Aveni, Richard, *The Pan-Industrial Revolution: How Manufacturing's Superheroes Are Changing the World?* translated by Ruxin Wang and Yanling Ye, Taipei: Aura Press, 2019.

Dai, J., and Tian, J. Tong, *A Study on the Regulation of Emerging Internet Industries*, Peking: China Development Press, 2017.

de Búrca, Gráinne, and Joanne Scott, eds, *New Governance and Constitutionalism in Europe and the US*, Oxford: Hart Publishing, 2006.

Derclaye, E., *The Legal Protection of Databases: A Comparative Analysis*, Edward Elgar Pub, 2008.

Detlor, B., Information Management, *International Journal of Information Management*, 2010, 30(2).

Diakopoulos, Nicholas, Rage Against the Algorithms, https://engenhonetwork.wordpress.com/2013/10/05/rage-against-the-algorithms-by-nicholas-diakopoulos/.

Domingoz, Pedro, *The Ultimate Algorithm: How Machine Learning and Artificial Intelligence Are Reshaping the World*, translated by Huang Fangping, Peking: CITIC Press, 2016.

Dorf, Michael C., After Bureaucracy, *The University of Chicago Law Review*, 2004, 71(3).

Dowdle, Michael W., Public Accountability: Conceptual, Historical, and Epistemic Mappings, in *Public Accountability: Designs, Dilemmas and Experiences*, ed. by Michael W. Dowdle, Cambridge: Cambridge University Press, 2006.

Downes, Anthony, *Inside the Bureaucracy*, translated by Guo, Xiaocong et al., Peking: People's University of China Press, 2006.

Elster, Jon, *Explaining Technical Change: A Case Study in the philosophy of Science*, Cambridge: Cambridge University Press, 1983.

Epidemic Watch, Digital Upgrade Research Report for Chinese SMEs (2020), http://www.cbdio.com/BigData/2020-04/13/content_6155587.htm.

Erkkilä, Tero, *Government Transparency: Impacts and Unintended Consequences*, Hampshire: Palgrave Macmillan, 2012.

Evans, Barbara J., Much ADO About Data Ownership, *Harvard Journal of Law and Technology*, 2011, 25(1).

Evans, Barbara J., Barbarians at the Gate: Consumer-Driven Health Data Commons and the Transformation of Citizen Science, *American Journal of Law and Medicine*, 2016, 42(4).

Evans, David, and Richard Schmalensee, *Matchmakers: The New Economics of Multisided Platforms*, New York: Harvard Business School Press, 2016.

Feng, Kaidong and Li, Junran, The Political Economy of Skills: Three Sets of Key Propositions, *Peking University Political Science Review* (4th series), Vol. 2, 2018.

Fernaeus, Ylva, Martin Jonsson, Jakob Tholander, *Revisiting the Jacquard Loom: Threads of History and Current Patterns in HCI*. CHI 2012, May 5–10, 2012, Austin, Texas, USA.

Fishkin, James S., *When the People Speak: Deliberative Democracy and Public Consultation*, Oxford: Oxford University Press, 2009.

Fogel, Robert William, *The Fourth Great Awakening and The Future of Egalitarianism*, Chicago: University Of Chicago Press, 2002.

Forde, Eamonn, Oversharing: How Napster Nearly Killed the Music Industry, https://www.theguardian.com/music/2019/may/31/napster-twenty-years-music-revolution.

Foucoult, M., *Discipline and Punish*, translated by A. Sheridan, Paris, FR, Gallimard, 1975.

Fountain, Jane E., Prospects for Improving the Regulatory Process Using: E-RULEMAKING, *Communications of the ACM. January*, vol. 46, no. 1, 2003.

Frey, Carl Benedikt, *The Technology Trap: Capital, Labor, and Power in the Age of Automation*, Princeton & Oxford: Princeton University Press, 2019.

Fung, Archon, Mary Graham, and David Weil, *Full Disclosure: The Perils and Promise of Transparency*, Cambridge: Cambridge University Press, 2007.

Gang, Phan, China's Wen Delivers Key Targets of its Development, *International Business Time*, Archived from the Original on 19 March 2011, Retrieved 5 March 2011.

Gascó-Hernández, Mila, ed., *Open Government: Opportunities and Challenges for Public Governance*, New York: Springer, 2014.

Gibson, Rachel K., Wainer Lusoli, Stephen Ward, Online Participation in the UK: Testing a 'Contextualised' Model of Internet Effects, *The British Journal of Politics & International Relations*, 2005, (4).

Glazer, R., Measuring the Value of Information: The Information-intensive Organization, *IBM Systems Journal*, 32(1), 1993.

Goldsmith, Jack and Tim Wu, *Who Controls the Internet: Illusions of a Borderless World*, Oxford: Oxford University Press, 2006.

Gordon, Robert, *The Rise and Fall of American Growth*, translated by Zhang Linshan et al., Peking: CITIC Publishing Group, 2018.

Gordon, Scott, *Controlling the State: A Constitutional History from Ancient Athens to Today*, translated by Ying Qi et al., Nanjing: Jiangsu People's Publishing House, 2005.

Gross, Miriam, *Farewell to the God of Plague: Chairman Mao's Campaign to Deworm China*, California: University of California Press, 2016.

Guoxin Securities, Overseas Market Week Overview: Mining the US Manufacturing Data, 5 January 2021, https://pdf.dfcfw.com/pdf/H3_AP2021010 61447750845_1.pdf.

Gurin, Joe, Big Data and Open Data: How Open Will the Future Be, *A Journal of Law and Policy*, 2015, 10, 691–704.

Gutman, Ami, Dennis Thompson, *Democracy and Disagreement*, translated by Yang Lifeng et al., Peking: Oriental Press, 2007.

Hall, Mark A., Property, Privacy and the Pursuit of Integrated Electronic Medical Records, *IOWA Law Review*, 2010, 95.

Hall, Peter A., and David Soskice, *Varieties of Capitalism: The Institutional Foundations of Comparative Advantage*, Oxford: Oxford University Press, 2001.

Han, Wenyan, Xiong, Yonglan and Zhang, Zhiqiang, Analysis of the Development Trend of the US Information and Communication Industry in the Last 20 Years and Implications, *World Science and Technology Research and Development*, 2021(2).

Harari, Yuval Noah, *A Brief History of Humankind*, translated by Lin, Junhong, Peking: CITIC Press, 2014.

Hogan, Mel, and Tamara Shepherd, Information Ownership and Materiality in an Age of Big Data Surveillance, *Journal of Information Policy*, 2015, 5.

Holmes, Stephen, Lineages of the Rule of Law, in *Democracy and the Rule of Law*, eds. by Jose Maria Maravall and Adam Przeworski, Cambridge: Cambridge University Press, 2003.

Houseman, Susan N., Understanding the Decline of U.S. Manufacturing Employment, Upjohn Institute Working Paper 18-287, https://research.upj ohn.org/cgi/viewcontent.cgi?article=1305&context=up_workingpapers.

Huang, Juan, and Huang, Zhuxiu, Big Data and Public Policy Research: Concepts, Relationships and Perspectives, *Chinese Administration*, 2015 (10).

Hughes, Eric, A Cypherpunk's Manifesto, http://www.activism.net/cypherpunk/manifesto.html.

Hutzel, Tim, and Dave Lippert, *Bringing Jobs Back to The USA: Rebuilding America's Manufacturing through Reshoring*, translated by He, Rong, Peking: People's Post and Telecommunications Publishing House, 2016.

Industrial Internet Talk, The Key to Cracking the Difficulty of Digital Transformation of SMEs is to Make Up for the Shortage of Talents, https://tech.china.com/article/20210918/092021_879389.html.

Jia, Kai, From 'Open Source Software' to 'Open Government': The Change of Government Governance Under the Influence of the Internet, *Comparative Economic and Social Systems*, 2016(2).

Jia, Kai, Global Governance of Cross-border Data Flows: Conflict of Power and Policy Cooperation: The Evolution of Regulatory Regimes for Cross-border Data Flows in Europe and the United States as an Example, *Journal of Shantou University* (Humanities and Social Sciences Edition), 2017(5).

Jia, Kai, Multiple Technology Routes for a Digital Future, *Culture Vertical*, no. 6, 2021.

Jia, Kai, and Yuhao Jiang, *The Technology and Politics of Digital Governance*, Peking: Social Science Literature Press, Forthcoming.

Jianjun, Dai and Tian Jietang, *Research on the Regulation of Emerging Internet Industries*, Peking: China Development Press, 2017.

Jing, Yijia ed., *The Road to Collaborative Governance in China*, New York: Palgrave Macmillan, 2015.

Kapczynski, Amy, The Law of Informational Capitalism, 129 *Yale Law Journal*, 2020, 1460.

Kaplan, Jerry, *The Age of Artificial Intelligence*, translated by Li, Peng, Peking: CITIC Press, 2016.

King, Ambrose Yeo-chi, A Voluntarist Model of Organization: The Maoist Version and Its Critique, *The British Journal of Sociology*, 1977, 28(3).

Kobielus, James, The Challenges of Transparent Accountability in Big Data Analytics, http://www.ibmbigdatahub.com/blog/challenges-transparent-accountability-big-data-analytics.

Koppell, Jonathan G. S., Pathologies of Accountability: ICANN and the Challenge of 'Multiple Accountabilities Disorder', *Public Administration Review*, 2005, 65(1), 94–108.

Kosack, Stephen, and Archon Fung, Does Transparency Improve Governance? *Annual Review of Political Science*, 2014, 17(2).

Kwok, D. W., *Scientism in Modern Chinese Thought (1900-1950)*, translated by Ji Lei, Nanjing: Jiangsu People's Publishing House, 1998.

Lai, Mao-Sheng, Gao Xin-Min, Zhao Xiao-Fan et al., Information Resources: New Factors of Production, Intangible Assets and Social Wealth, *China Informatization*, 2008(10).

Lazonick, William, *Competitive Advantage on the Shop Floor*, Cambridge: Harvard University Press, 1990.

Lazonick, William, *Sustainable Prosperity in the New Economy: Business Organization and High-Tech Employment in the United States*, W. E. Upjohn Institute, 2009.

Lehr, William, and David Clark, Steve Bauer, Arthur Berger and Philipp Richter, Whither the Public Internet? *Journal of Information Policy*, 2019, 9.

Li, Lin, and Liu, Yu, A Comparative Analysis of Fair Value Measurement Model and Historical Cost Measurement Model, *Business Economics*, 2008(01).

Li, Zehou, *Practical Reason and the Culture of Joy*, Peking: Sanlian Bookstore, 2005.

Lin, Justin Yifu, *Interpreting the Chinese Economy*, Peking: Peking University Press, 2012.

Lipsey, Gichard G., Kenneth I. Carlaw, and Clifford T. Bekar, *Economic Transformations: General Purpose Technologies and Long Term Economic Growth*, New York: Oxford University Press. New York: Oxford University Press, 2005.

Lord, Kristin M., *The Perils and Promise of Global Transparency: Why the Information Revolution May Not Lead to Security, Democracy, or Peace*, New York: State University of New York Press, 2006.

Lothian, Tamara, *Law and the Wealth of Nations: Finance, Prosperity and Democracy*, New York: Columbia University Press, 2017.

Lu, Xun, On Cultural Bias to, in *The Complete Works of Lu, Xun* (vol. 1), Peking: People's Literature Publishing House, 2005.

Lukács, G., *The Destruction of Reason*, translated by Wang Jiuxing, Nanjing: Jiangsu Education Press, 2005.

Lv, Xinyu, and Zhao, Yuezhi, Modernity, Mass Media and the Reconstruction of Publicness in China, in *Love Thought website*, http://www.aisixiang.com/data/34285.html.

Ma, Dung, Yu, Xin, Data Assets: Conceptual Evolution and Measurement Methods, *Journal of Statistics*, 2020(02), 15–24.

Mao, Zedong, Transforming Our Learning, in *Selected Works of Mao, Zedong* (vol. 3), Peking: People's Publishing House, 1991.

Marx, Karl, Capital, vol. 1, in *Marx and Engels Collected Works*, vol. 5, translated by the Bureau for the Compilation of the Works of Marx, Engels, Lenin and Stalin of the Central Committee of the Communist Party of China, Peking: People's Publishing House, 2009.

Mazzucato, Mariana, *The Value of Everything: Making and Taking in the Global Economy*, London: Penguin, 2019.

Mazzucato, Mariana, *Innovative Government: Building a Symbiotic and Win-win Relationship Between Public and Private Sectors*, translated by Lei, Li, Dongxin Shuang and Shanjian Cheng, Peking: CITIC Publishing House, 2019.

McCarthy, J., *What is Artificial Intelligence*, http://www-formal.stanford.edu/jmc/whatisai/whatisai.html.

McKinsey Global Institute, *Digital America: A tale of the haves and have-mores*, December 1, 2015, https://www.mckinsey.com/industries/technology-media-and-telecommunications/our-insights/digital-america-a-tale-of-the-haves-and-have-mores.

Milanovic, Branko, *Global Inequality: A New Approach for the Age of Globalization*, Cambridge, Massachusetts and London: The Belknap Press of Harvard University, 2016.

Milanovic, Branko, *Capitalism, Aone: The Future of the System That Rules the World, Cambridge*, Massachusetts and London: The Belknap Press of Harvard University, 2019.

Mingers, J., *Realizing Systems Thinking: Knowledge and Action in Management Science*, Berlin: Springer, 2006.

Mo, Jiwu, Yang, Changming, and Jiang, Yuhao, *Consultative Democracy and Orderly Participation*, Peking: China Social Science Press, 2009.

Morabito, Vincenzo, *Business Innovation Through Blockchain: The B^3 Perspective*, Springer, 2017.

Mu, Yong, *Top-level Design Framework and Application Scheme for Peking's Big Data Platform for Government Affairs*, Available from http://www.haokoo.com/internet/1840119.html.

Mullaney, Thomas S., *The Chinese Typewriter: A History*, Cambridge: The MIT Press, 2017.

Mumford, Lewis, *Technology and Civilization*, translated by Chen Yunming, Wang, Keren and Li, Huashan, Peking: China Construction Industry Press, 2009.

Negroponte, Nicholas, *A Special Preface to the 20th Anniversary Chinese Edition of Digital Survival*, translated by Fan Haiyan, in Future World: http://www.51wxjz.com/100473.

Noble, David F., *Forces of Production: A Social History of Industrial Automation*, New Brunswick and London: Transaction Publishers, 2011.

Novick, Peter, *The Holocaust and Collective Memory*, translated by Wang Zhihua, Nanjing: Yilin Publishing House, 2019.

O'Mara, Margaret, *The Silicon Valley Code: How Technological Innovation is Reshaping America*, translated by Xie, Nijie, Peking: CITIC Publishing House, 2022.

Ou, Shujun, *The Foundation of National Basic Capacity*, Peking: China Social Science Press, 2013.

Palmer M., Data is the New Oil, http://ana.blogs.com/maestros/2006/11/data_is_the_new.html.

Parker, Geoffrey G., Marshall W. Van Alstyne, and Sangeet Paul Choudary, *Platform Revolution: How Networked Markets are Transforming the Economy—and How to Make Them Work for You*, New York and London: W. W. Norton & Company, 2016, 4.

Pasquale, Frank, *The Black Box Society: The Secret Algorithms That Control Money and Information*, Cambridge: Harvard University Press, 2006.

Piketty, Thomas, *Capital and Ideology*, Cambridge, Massachusetts and London: The Belknap Press of Harvard University, 2020.

Proudhon, J., *The Philosophy of Poverty* (vol. 1), translated by Yu Shutong and Wang Xuehua, Peking: The Commercial Press, 2010.

Qiu, Zeqi et al., From the Digital Divide to Dividend Differences—A Perspective on Internet Capital, *Chinese Social Sciences*, 2016, 10.

Qiu, Zeqi and Qiao Tianyu, Technological Changes in E-Commerce and Farmers' Co-Development, *Chinese Social Sciences*, 2021, 10.

Qu, Qiubai, Experimentalism and Revolutionary Philosophy, in *Selected Writings of Qu Qiubai*, Chengdu: Sichuan Literature and Art Publishing House, 2010.

Ran, Ran, Political Incentives and Local Environmental Governance in a 'Pressure-Based System', *Comparative Economic and Social Systems*, 2013(3).

Rasmussen, Terje, Internet and the Political Public Sphere, *Sociology Compass*, December 2014, 8(12).

Raymond, Eric, The Cathedral and the Bazaar. *Knowledge, Technology & Policy*, 1999, 12, 23–49, https://doi.org/10.1007/s12130-999-1026-0.

Rees, Lawrence, *Auschwitz: The Nazis & the 'Final Solution' (Auschwitz: A New History)*, translated by Liu Shuang, Nanjing: Yilin Publishing House, 2020.

Rochet, Jean-Charles, and Jean Tirole, Platform Competition in Two-Sided Markets, *Journal of the European Economic Association*, June 2003, 1(4), 990–1029.

Rock, Michael T., and Michael A. Toman, *China's Technological Catch-up Strategy: Industrial Development, Energy Efficiency, and* CO_2 *Emissions*, Oxford: Oxford University Press, 2015.

Rodirik, Dani, New Technologies, Global Value Chains, and the Developing Economies, NBER Working Paper Series, October 2018, http://www.nber.org/papers/w25164.

Rodwin, Marc A., Patient Data: Property, Privacy & the Public Interest, *American Journal of Law & Medicine*, 2010, 36.

Rong, Jingben, Cui Zhiyuan et al., *The Shift from a Pressure-Based System to a Democratic and Cooperative System*, Peking: Central Compilation Press, 1998.

Rosenbach, Marcel, How Google and Facebook Can Reshape Elections, http://www.spiegel.de/international/germany/google-and-facebook-could-help-decide-2017-german-election-a-1120156.html.

Sabel, Charles, Dewey, Democracy, and Democratic Experimentalism, *Contemporary Pragmatism*, December 2012, 9(2).

Sabel, Charles F., and William H. Simon, Contextualizing Regimes: Institutionalization as a Response to the Limits of Interpretation and Policy Engineering, *Michigan Law Review*, May 2012, 110(7).

Sabel, Charles F., and William H. Simon, The Duty of Responsible Administration and the Problem of Police Accountability, *Yale Journal on Regulation*, 2016, 33(1), 165–211.

Sabel, Charles F., and Jonathan Zeitlin, Experimentalist Governance, in David Levi-Faur ed., *The Oxford Handbook of Governance*, Oxford: Oxford University Press, 2012.

Sandulli, Francesco D., Solving the Paradoxes of the Information Technology Revolution: Productivity and Inequality, https://papers.ssrn.com/sol3/papers.cfm?abstract_id=1192568.

Schmitt, Carl, The Age of Neutrality and Depoliticisation, in *Schmitt's Collected Works: The Concept of Politics*, translated by Zongkun Liu et al., Shanghai: Shanghai People's Publishing House, 2004.

Scott, Colin, Accountability in the Regulatory State, *Journal of Law and Society*, March 2000, 27(1), Voices, Spaces, and Processes in Constitutionalism.

Selyukh, Alina, Section 230: A Key Legal Shield For Facebook, Google is About to Change, https://www.npr.org/sections/alltechconsidered/2018/03/21/591622450/section-230-a-key-legal-shield-for-facebook-google-is-about-to-change.

Shapiro, Karl, Hal Van Rijn, *Information Rules: A Strategic Guide to the Network Economy*, translated by Meng, Zhaoli and Niu, Luqing, Peking: Renmin University of China Press, 2013.

Shkabatur, Jennifer, Transparency With(out) Accountability: Open Government in the United States, *Yale Law & Policy Review*, 2012, 31(1).

Smythe, Dallas, What Comes After the Bicycle?: The Politics and Ideology of Technology, translated by Wang, Hongzhe, *Open Times*, no. 4, 2014.

Solow, Robert, We'd Better Watch Out, *New York Times Book Review*, July 12, 1987.

Srnicek, N. & Williams, A. The Future isn't Working, *Juncture*, 2015, 22(3).

Standing, G., *The Precariat: The New Dangerous Class*, London: Bloomsbury Publishing Plc, 2011.

Stiglitz, J. E., Macroeconomic Fluctuations, Inequality, and Human Development, *Journal of Human Development and Capabilities*, 2012.

Swan, Melanie, *Blockchain: Blueprint for a New Economy*, O'Reilly Media, 2015.

Tencent Research Institute, Twenty Years of Internet Regulation in China, https://www.tisi.org/15948.

Thaler, Richard H., Cass R. Sunstein, *Nudge: Improving Decisions About Health, Wealth, and Happiness*, New Haven & London: Yale University Press, 2008.

Thierer, A. D., and Watney, C. J., Comment on the Federal Automated Vehicles Policy, 2016.

The Commentator, Playing a Good First Move, Playing a Good Active Battle—Overview of General Secretary Xi Jinping's Important Discussion on the Prevention and Resolution of Major Risks, *People's Daily*, 15 April 2021.

Thelen, Kathleen, *How Institutions Evolve: The Political Economy of Skills in Germany, Britain, the United States, and Japan*, New York: Cambridge University Press, 2004.

Thierer, A., *Permissionless Innovation: The Continuing Case for Comprehensive Technological Freedom*, Mercatus Center at George Mason University, 2016.

Toffler, Alvin, *The Impact of the Future*, translated by Cai, Xinzhang, Peking: CITIC Press, 2006.

Toulmin, Stephen, *Cosmopolis: Hidden Agenda of Modernity*, Chicago: University of Chicago Press, 1990.

Tucker, Ian, Interview: Evgeny Morozov: 'We are Abandoning all the Checks and Balances', https://www.theguardian.com/technology/2013/mar/09/evgeny-morozov-technology-solutionism-interview.

Turing, A. M., Computing Machinery and Intelligence, *Mind*, 1950, 59(236).

Tyson, Laura, Michael Spencer, The Impact of Technology on Income and Wealth Inequality, in Heather Bussey, Bradford DeLong, Marshall Steinbaum, eds, *After Piketty: A New Agenda for Inequality Research*, translated by Yu Jiang and Gao Desheng, Peking: CITIC Press, 2022.

Unger, Roberto M., *False Necessity: Anti-Necessitarian Social Theory in the Service of Radical Democracy*, London and New York: Verso, 2004.

Unger, Roberto M., *The Knowledge Economy*, London and New York: Verso, 2019.

Unger, Roberto M., The Knowledge Economy: A Critique of the Dominant View, *American Affair*, Fall 2020/Volume IV, Number 3.

Unger, R. M., Isaac Stanley, Madeleine Gabriel and Geoff Mulgan, *Imagination Unleashed: Democratising the Knowledge Economy*, Published by Nesta, March 2019.

Van Parijs, P., Basic Income: A Simple and Powerful Idea for the Twenty-first Century, *Politics & Society*, 2004, 32(1).

Wang, Hongzhe, From 'Barefoot Electrician' to 'Electronic Bagman': Technology and Labour Politics in China's Electronic Information Industry, *Open Times*, no. 3, 2015.

Wang, Hui, *The Emergence of Modern Chinese Thought* (vol. 2), Peking: Sanlian Bookstore, 2008.

Wang, Yang, An Analysis of the Ownership of Securities Trading Information and its Legal Protection Model—From the Winful Data Case, *Special Economic Zone*, 2013(4).

Ward, Jonathan Stuart, and Adam Barker, Undefined By Data: A Survey of Big Data Definitions, https://arxiv.org/pdf/1309.5821.pdf.

Weber, Max, *The Sociology of Domination*, translated by Kang Le and Jian Hhuimei, Guilin: Guangxi Normal University Press, 2004.

Weng, Leyen, and Li, You-Yun, Openness and Sharing of Big Data in Government Affairs: A Study of Conditions, Obstacles and Basic Guidelines, *Comparative Economic and Social Systems*, 2016(2), 113–122.

Womack, James P., Daniel T. Jones and Daniel Roos, *The Machine That Changed the World*, New York: Free Press, 1990.

Wood, B. Dan, and Richard W. Waterman, The Dynamics of Political Control of the Bureaucracy, *The American Political Science Review*, 1991, 85(3).

Wu, Shenguo, The Global Situation of Data Governance and China's Response Strategy, *E-Government*, 2019(1).

Xi, Jinping, Speech at the Opening Ceremony of the Second World Internet Conference (full text), in Xinhua, http://www.xinhuanet.com/politics/2015-12/16/c_1117481089.htm.

Xi, Jinping, Solidly Promoting Common Prosperity, *Seeking Truth*, no. 20, 2021. Available from Seeking Truth, http://www.qstheory.cn/dukan/qs/2021-10/15/c_1127959365.htm.

Xi, Jinping, Keep Making China's Digital Economy Stronger, Better and Bigger, *Seeking Truth*, 2022, no. 2. In Seeking Truth, http://www.qstheory.cn/dukan/qs/2022-01/15/c_1128261632.htm.

Xing, Yuqing, *The Enigma of Chinese Exports: Decoding the 'Global Value Chain'*, Peking: Sanlian Bookstore, 2022.

Xue, Liang, Japan Pushes to Realize Super-smart Society 'Society 5.0', http://www.istis.sh.cn/list/list.aspx?id=10535.

Ye, Jiyuan, Chen, Meng, Xie Huan et al., An Exploration of the Logical Relationship Between Data and Information—and the DIKW Concept Chain Model, *Chinese Journal of Library Science*, 2017(03).

Zhang, Huiping, and Ma, Taiping, Cross-Border Flow of Data Elements in Comprehensive Digital Transformation of Cities: Four Models, Promotion Logic and Innovation Paths, *e-Government*, no. 5, 2022.

Zhang, Nan, Public-Derived Big Data Analysis and Government Decision-Making Process Reconstruction: Theoretical Evolution and Research Outlook, *Chinese Administration*, 2015(10).

Zhang, Peng, and Jiang, Yuhao, Basic Theoretical Research on Asset-Based Management of Government Data: Asset Attributes, Data Ownership and Pricing Methods, *e-Government*, no. 9, 2020.

Zhang, Weiying, and Liu He, eds., *Research Report on E-Government in China's Prefectural Cities*, Peking: Economic Science Press, 2003.

Zheng, Yongnian, *Technological Empowerment: The Internet, State, and Society in China*, California: Stanford University Press, 2008.

Zheng, Yongnian, China and the U.S.: A Clash of Techno-Nationalists, *The Washington Post*, May 15, 2018.

Zhou, Hanhua, Exploring the Problems and Countermeasures of the Implementation of the Regulations on Open Government Information, *China Administration*, no. 7, 2009.

Zhou, H., Building the Rule of Law in Cyberspace, http://news.youth.cn/jsxw/201410/t20141026_5910191.htm.

Zhou, Yanbing, The Development and Application of Big Data Industry in France, *Study Times*, 2014-11-03.

Zizek, Slavoj, We Need a Socialist Reset, Not a Corporate 'Great Reset', https://jacobinmag.com/2020/12/slavoj-zizek-socialism-great-reset.

Zou, Tang, *The Cultural Revolution and Post-Mao Reforms: A Historical Perspective*, Chicago and London: The University of Chicago Press, 1986.

Zuo, Xiaodong, Exploring Some Technical Issues in the Network Security Management of Cloud Computing Services, GlobeNewswire: http://world.huanqiu.com/hot/2015-07/7135068.html.

Zuo, Yilu, *Beyond the 'Street Corner Speaker': The Margins and the Centre of the Right to Expression*, Peking: Social Science Literature Press, 2020.

Zuboff, Shoshana, *The Age of Surveillance Capitalism, The Fight for a Human Future at the New Frontier of Power*, New York: Public Affairs, 2019.

Analysis of the Size and Growth Prospects of the Online Car Market in 2020, in Foresight Economist, https://baijiahao.baidu.com/s?id=1667000376138315188&wfr=spider&for=pc.

Artificial Intelligence: Opportunities and Implications for the Future of Decision Making, https://www.gov.uk/government/uploads/system/uploads/attachment_data/file/566075/gs-16-19-artificial-intelligence-ai-report.pdf.

Bonina, C. M., New Business Model and the Value of Open Data, http://pdfs.semanticscholar.org/ceaf/d56a1172833a74246f4f3084b17d5e8b54af.pdf.

CNNIC, The 43rd Statistical Report on the Development Status of the Internet in China 2019, http://www.199it.com/archives/839540.html.

Elon Musk, Artificial Intelligence is Our Biggest Existential Threat, https://www.theguardian.com/technology/2014/oct/27/elon-musk-artificial-intelligence-ai-biggest-existential-threat.

How the IBM PC Won, Then Lost, the Personal Computer Market, https://spectrum.ieee.org/how-the-ibm-pc-won-then-lost-the-personal-computer-market.

Inflated by 575 times in 11 Years, Why are Mobile App Installers Getting Bigger? https://mp.weixin.qq.com/s/1XBhcnDC0Ahm-QYzmNqYxA.

Microsoft's Bill Gates Insists AI is A Threat, http://www.bbc.com/news/31047780.

Notice of the State Council on the Issuance of the Development Plan for a New Generation of Artificial Intelligence, http://www.gov.cn/zhengce/content/2017-07/20/content_5211996.htm.

Preparing for the Future of Artificial Intelligence, https://obamawhitehouse.archives.gov/sites/default/files/whitehouse_files/microsites/ostp/NSTC/preparing_for_the_future_of_ai.pdf.

Stephen Hawking says A.I. could be 'Worst Event in the History of Our Civilization', https://www.cnbc.com/2017/11/06/stephen-hawking-ai-could-be-worst-event-in-civilization.html.

The Consumer Electronics Hall of Fame, Tandy/RadioShack Trs-80, https://spectrum.ieee.org/the-consumer-electronics-hall-of-fame-tandyradioshack-trs80-model-1.

The Internet Society of China, History of Internet Development in China (Chronology), https://www.isc.org.cn/ihf/info.php?cid=218.

The President in Conversation With MIT's Joi Ito and WIRED's Scott Dadich, https://www.wired.com/2016/10/president-obama-mit-joi-ito-interview/.

The Rise of the Wintel Empire, https://abcnews.go.com/Technology/story?id=98348&page=1.

Windows + Intel = Wintel? https://www.mayin.org/ajayshah/MEDIA/1998/wintel.html.

Index

Y. Jiang and K. Jia, *Transformation of Platform Governance in China*, IPP Studies in the Frontiers of China's Public Policy,
https://doi.org/10.1007/978-981-99-6456-7

GPSR Compliance

The European Union's (EU) General Product Safety Regulation (GPSR) is a set of rules that requires consumer products to be safe and our obligations to ensure this.

If you have any concerns about our products, you can contact us on ProductSafety@springernature.com

In case Publisher is established outside the EU, the EU authorized representative is:

Springer Nature Customer Service Center GmbH
Europaplatz 3
69115 Heidelberg, Germany

FSC
www.fsc.org
MIX
Papier | Fördert gute Waldnutzung
FSC® C083411

Kolibri 360 GmbH
Ferdinand-Jühlke-Straße 7
99095 Erfurt, Deutschland
produktsicherheit@kolibri360.de